ADVANCES IN SYSTEMS RESEARCH SERIES
Volume 1 - *Editor:* J. M. GVISHIANI

SYSTEMS RESEARCH
Methodological Problems

Other Pergamon Titles of Interest

AKASHI	Control Science & Technology for the Progress of Society, 7 volumes
BUZACOTT	Scale in Production Systems
ELLOY & PIASCO	Classical & Modern Control with Worked Examples
EYKHOFF	Trends and Progress in System Identification
FRONZA & MELLI	Mathematical Models for Planning & Controlling Air Quality
ISERMANN	System Identification Tutorials
LASKER	Applied Systems & Cybernetics, 6 volumes
MAHMOUD & SINGH	Large Scale Systems Modelling
MILLER	Distributed Computer Control Systems, 1981
MORRIS	Communication for Command & Control Systems
NAJIM	Systems Approach for Development
PATEL & MUNRO	Multivariable Systems Theory & Design
SINGH et al.	Applied Industrial Control. An Introduction
SINGH & TITLI	Systems: Decomposition, Optimisation & Control
TITLI & SINGH	Large Scale Systems: Theory and Applications
TZAFESTAS	Distributed Parameter Control Systems

Pergamon Related Journals

(Free Specimen Copies Gladly Sent on Request)

AUTOMATICA
COMPUTERS & ELECTRICAL ENGINEERING
COMPUTERS & INDUSTRIAL ENGINEERING
COMPUTERS & MATHEMATICS with Applications
COMPUTERS & OPERATIONS RESEARCH
INFORMATION SYSTEMS
INFORMATION PROCESSING & MANAGEMENT
SYSTEMS RESEARCH

SYSTEMS RESEARCH
Methodological Problems

Prepared by
USSR State Committee for Science and Technology
USSR Academy of Sciences
Institute for Systems Studies

Translated by
E. L. NAPPELBAUM
YU. A. YAROSHEVSKII
and
D. G. ZAYDIN

PERGAMON PRESS
OXFORD · NEW YORK · TORONTO · SYDNEY · PARIS · FRANKFURT

U.K.	Pergamon Press Ltd., Headington Hill Hall, Oxford OX3 0BW, England
U.S.A.	Pergamon Press Inc., Maxwell House, Fairview Park, Elmsford, New York 10523, U.S.A.
CANADA	Pergamon Press Canada Ltd., Suite 104, 150 Consumers Road, Willowdale, Ontario M2J 1P9, Canada
AUSTRALIA	Pergamon Press (Aust.) Pty. Ltd., P.O. Box 544, Potts Point, N.S.W. 2011, Australia
FRANCE	Pergamon Press SARL, 24 rue des Ecoles, 75240 Paris, Cedex 05, France
FEDERAL REPUBLIC OF GERMANY	Pergamon Press GmbH, Hammerweg 6, D-6242 Kronberg-Taunus, Federal Republic of Germany

Copyright © 1984 Pergamon Press Ltd.

All Rights Reserved. No part of this publication may be reproduced, stored in a retrieval system or transmitted in any form or by any means: electronic, electrostatic, magnetic tape, mechanical, photocopying, recording or otherwise, without permission in writing from the publishers.

First edition 1984

Library of Congress Cataloging in Publication Data
Gvishiani, Dzhermen Mikhailovich.
Systems research.
1. System theory. 2. System analysis. I. Title.
Q295.G9413   1983   003   82-22583

British Library Cataloguing in Publication Data
Systems research.
1. System theory.
I. Gvishiani, J. M.
003   Q295
ISBN 0-08-030000-6

In order to make this volume available as economically and as rapidly as possible the authors' typescripts have been reproduced in their original forms. This method unfortunately has its typographical limitations but it is hoped that they in no way distract the reader.

Printed in Great Britain by A. Wheaton & Co. Ltd., Exeter

EDITORIAL BOARD

J. M. GVISHIANI (Editor-in-Chief)
V. N. SADOVSKY (Associated Editor)
V. L. ARLAZAROV, B. V. BIRIUKOV, I. V. BLAUBERG,
V. I. DANILOV-DANILYAN, K. M. KHAILOV, O. A. KOSSOV,
N. I. LAPIN, O. I. LARICHEV, V. A. LEKTORSKY,
A. A. MALINOVSKY, E. S. MARKARIAN, B. Z. MILNER,
E. M. MIRSKY, E. L. NAPPELBAUM, I. B. NOVIK,
D. A. POSPELOV, A. I. UEMOV, B. G. YUDIN, V. P. ZINCHENKO

PREFACE

With this volume Pergamon Press begins publication of selected papers from *Systems Research: Methodological Problems*, a yearbook which has been published in Moscow by Nauka Publishing House since 1969. Being very similar to the General Systems Yearbook which has been published in the United States since 1956, *Systems Research: Methodological Problems* is a serial publication which offers comprehensive treatment of problems of systems research.

In the USSR research into this field has been actively pursued for over 20 years. The first Soviet publications on the methodology of systems analysis appeared almost simultaneously in philosophical, technical and biological literature. Today there is hardly a sphere of science and technology that does not employ the principles of wholeness, structural and hierarchical organization, and other basic methodological principles of the systems approach. In the 1970s, research in the modelling of processes of global development, and investigations associated with the solution of large-scale social, economic, management and other problems gave a powerful impetus to the development of this field of knowledge.

The yearbook *Systems Research: Methodological Problems* is expected to play a part in the development of modern systems research, as its main functions are to make practical use of the methodological possibilities of the systems approach and to expand the area of their application. Between 1969 and 1978 the yearbook was published by the Institute for History of Science and Technology, USSR Academy of Sciences and since 1979 it has been brought out by the Institute for Systems Studies, of the State Committee for Science and Technology and the USSR Academy of Sciences. The editors believe that today priority should be given to four fundamental areas of research: to identify and develop general philosophical principles of systems analysis, to study its interdisciplinary character, to evolve the conceptual apparatus of systems analysis and general systems theory, and, finally, to devise ways and means of applying the systems approach to specific practical and scientific problems. In this context the yearbook *Systems Research: Methodological Problems* is expected to pool the experience of those who do theoretical and practical work in different fields of science and technology using systems analysis.

Preface

This book includes publications which are of interest from the methodological point of view and which appeared in the Yearbook at different times between 1969 and 1979.

The articles in the first section are primarily concerned with the formulation of general systems theory and systems analysis as methodological concepts which may be used to study entities in their evolution. In this section a great deal of attention is devoted to the application of the methodological tools of systems analysis and general systems theory to the study of economic and scientific problems which has been a standing feature of the Yearbook over a long period of time.

The articles in the second section deal with the development of substantive and mathematical models of different patterns of social and economic behaviour. The authors of these articles analyse the basic laws governing the operation of the mechanisms which control human behaviour. The latter is viewed as a hierarchically organized system the different levels of which interact in a wide variety of ways.

The third section deals with various methodological problems of systems analysis and mathematical modelling of scientific activity, with the emphasis placed on the performance of communication mechanisms which regulate the generation and dissemination of scientific knowledge. The authors of the articles in this section concentrate primarily on the establishment and examination of the factors and conditions which support effective communication between researchers (or those responsible for other functions) and consumers of scientific knowledge. Also explored in this context is the relationship between the functioning of communication mechanisms in science and its rate of progress.

Currently the publication of the second volume of selected papers from *Systems Research: Methodological Problems 1980-1982* is at the stage of preparation.

The editors hope that the publication of selected papers from *Systems Research: Methodological Problems* in English will stimulate progress in the development and practical application of the methodology of systems analysis and will promote understanding of its role in modern science.

The Editors

CONTENTS

METHODOLOGICAL PROBLEMS OF SYSTEMS APPROACH AND SYSTEMS ANALYSIS

J. M. Gvishiani. Materialist Dialectics as a Philosophical Basis for Systems Research.   3

V. N. Sadovsky. Systems Approach and General Systems Theory — State of the Art, Main Problems and Prospects for Development.   16

E. L. Nappelbaum. Systems Analysis as a Research Programme: Guidelines and Objectives.   33

I. V. Blauberg. System and Wholeness Concepts.   49

E. G. Yudin. Activity and Systems Approach.   62

A. A. Malinovsky. Basic Concepts and Definitions in Systems Theory as Applied to Biology.   82

K. M. Khailov. Principles of Systems Research and Applied Biology.   90

G. A. Smirnov. The Underlying Concepts of the Formal Theory of Wholeness.   98

MODELLING METHODS IN SYSTEMS ANALYSIS

N. I. Lapin. Nonformalised Components of Modelling Systems.   115

Yu. A. Levada. A Model of the Reproductive System: Conceptual Categories.   127

V. I. Danilov-Danilyan, I. L. Tolmachev, V. V. Shurshalov. On the Problem of Simulation of Systems with Changing Structure.   134

O. I. Larichev. Methodological Aspects of Practical Applications of Systems Analysis.   146

N. F. Naumova. A Systems Description of Goal-seeking Human Behaviour.   154

SYSTEMS APPROACH TO THE SCIENCE OF SCIENCE

E. M. Mirsky. Body of Publications and Science Discipline System. 171

A. A. Ignatiev. The Holistic Principle in an Interdisciplinary Study of Scientific Activity. 192

A. I. Yablonsky. The Development of Science as an Open System. 211

A. P. Ogurtsov. Disciplinary Knowledge and Scientific Communications. 229

S. I. Doroshenko. Scientometrological Indicators of Soviet Literature of Systems Research. 246

Subject Index 253

Methodological Problems of Systems Approach and Systems Analysis

MATERIALIST DIALECTICS AS A PHILOSOPHICAL BASIS FOR SYSTEMS RESEARCH

J. M. Gvishiani

Through constant expansion of its social role science has become a major tool for controlling the development of human society. While exercising this function science itself undergoes quantitative and qualitative changes which follow two trends: differentiation and integration of its various domains. L. I. Brezhnev pointed out that "fresh opportunities for fruitful general theoretical, fundamental and applied research arise in interdisciplinary areas, notably in the natural and the social sciences" [3].

The integrative trend appears at its most productive in systems research, which is a vigorously developing area with only a few decades to its history. Systems research treats complex, multiple, large-scale problems (including global socioeconomic ones) and consistently seeks not only to get to the heart of the matter but to devise ways and means for a rational control of the objects in question and to promote the solution of the problems on hand. This unity of study and transformation is responsible for the comprehensive interdisciplinary nature of systems research.

In the Soviet Union and other Socialist countries systems research is founded on the philosophical basis of materialist dialectics. This paper offers a substantiation of this conception.

DIALECTICS AND THE PHILOSOPHIC SYSTEMS PRINCIPLE

Before tackling the main subject of this paper I shall first introduce some definitions. A system in the most general sense is defined as *a complex of interrelated elements forming a whole*. This understanding of a system imposes no constraints on the nature of the constituent elements (they may be material objects or ideal structures) or the kind of whole formed (it may vary from a mechanical sum of externally connected material objects to, in Marx's words, an organic integrity like that of a living creature, the human brain, the organized structure of a large industrial enterprise, a social structure). The main task of systems research lies in the analysis and construction of various systems and the control of natural systems. It is essential to distinguish between systems as objects of study, construction of problems inherent in an object system or important to the subject (researcher).

Systems research is a rapidly developing sphere of science and technology and it requires a philosophic substantiation. Of fundamental importance here is the philosophic systems principle which asserts that a "a phenomenon of objective reality viewed in terms of a systematic whole and the interaction of its constituent elements is a specific cognitive prism or a specific 'dimension' of reality" [17, p. 10].

Apart from its concrete scientific value, the systems principle has a broad conceptual significance; it is the upshot of a long evolution of natural science and philosophy.

The principle itself first came to be understood within the framework of 16th century natural philosophy which strove to "systematize" the knowledge about nature obtained by concrete sciences. In reality, however, natural philosophy did not go further than a quasi-systematic approach which consisted in a fixed classification or a metaphysical arrangement of all things within an unchangeable universal "system". Along with the first realization of the systems principle, this was, in fact, the first manifestation of the tendency to use the term "system" loosely to describe any more or less organized totality of objects.

Schelling and then Hegel put in a great deal of effort to join "systemization" to development. However, the idealist conception of the natural and social world as a reflection of an absolute idea led both thinkers to enclose the systems principle inside philosophy itself, which left the dialectic principle of development crushed in the Procrustean bed of metaphysics.

Nevertheless classic German philosophy made a big step forward by demonstrating that object-knowledge or knowledge about objects *per se* is but direct, simple knowledge which must serve to attain essential, substantial or systematic knowledge that treats an object as part of a broader system. This idea was clearly laid down by Hegel in his *Encyclopaedia of Philosophic Sciences*:

> Unless it is a system, a philosophy is not a scientific production. Unsystematic philosophising can only be expected to give expression to personal peculiarities of mind, and has no principle for the regulation of its contents. Apart from the interdependence and organic union, the truths of philosophy are valueless, and must be treated as baseless hypotheses, or personal convictions. Yet many philosophical treatises confine themselves to such an exposition of the opinions and sentiments of the author [11, vol. 1, p. 100].

The systems principle found its first truly scientific expression in the works of Marx and Engels. The dialectic materialist treatment of the natural and social world freed the systems principle as well as the principle of development from intraphilosophic confinement. This is how Engels characterized the unity of these dialectic principles as applied to material reality in *Dialectics of Nature*:

> The whole of nature accessible to us forms a system, an interconnected totality of bodies, and by bodies we understand here all material existences extending from stars to atoms .... In the fact that these bodies are interconnected is already included that they react on one another, and it is precisely this mutual reaction that constitutes motion. It already becomes evident here that matter is unthinkable without motion. And if, in addition, matter confronts us as something given, equally uncreatable as indestructible, it follows that motion also is as

uncreatable as indestructible. It became impossible to reject this
conclusion as soon as it was recognized that the universe is a system,
an inter-connection of bodies [1, vol, 20, p. 392].

In *Capital* Marx applied the systems principle in organic unity with the
principle of development to the analysis of a concrete socio-economic system
and used it to create a theory of that system. This application bore rich
fruit for the systems principle itself: systems properties were shown to
exist as integral properties of a system; social phenomena were found to be
peculiar in that man doubles himself in the things he creates and the nature
he transforms, i.e. natural qualities are supplemented with man-made social
qualities in the "second nature"; there is a duality to the social qualities
themselves, as in the dual nature of labour. The evolution of mankind is
regarded by the Marxist theory as onward motion from systems with a
predominance of natural determination, and within the latter category from
an economic entity to a higher truly social community. Marx's elaboration
of the systems principle yielded the fundamental conclusion that real objects
in general and social phenomena in particular belong to several systems
instead of one. Hence any real, i.e. truly concrete, knowledge is
polysystemic instead of merely systemic.*

Thus in materialist dialectics the systems principle is joined organically
to the principle of development and performs methodological functions with
regard to both ontology and the theory of cognition. To discover the
polysystematic nature of scientific knowledge one needed an understanding of
the infinity and inexhaustibility of systems relations in the objective
material world. Therefore one is fully justified in saying that "the
systems principle is one of the major components of marxist theory and
methodology" [17, p. 20]. The subsequent development of philosophy and
special sciences has fully confirmed the materialist dialectic
interpretation of the systems principle.

The crisis that struck the methodology of natural science at the turn of the
twentieth century and was profoundly analysed by V. I. Lenin in *Materialism
and Empiriocriticism* [2] led most naturalists to renounce the mechanistic
conception of the world as a sum of separate unchangeable elements and accept
the dialectical view of the world as an evolving systematic whole. The rapid
development of science in the twentieth century not only deepened and refined
the philosophic notions and principles of scientific cognition but altered
and enriched the entire machinery of scienctific categories,leading *inter
alia*, to a greater significance of the general scientific concepts and the
elaboration of general scientific methodological principles and approaches.
It is in this sphere that the struggle between materialism and idealism in
scientific cognition came to be centred: the neopositivists regarded the
general scientific approaches and concepts as a substitute for philosophy
while the materialists saw them as a new proof of the efficiency of
materialist dialectics, a new expedient for an adequate reflection of reality
and its more effective transformation by man. The new general scientific

---

*For detailed treatment of this and other aspects of systems thinking in
Marx's doctrine see *The Systems Principle in Marx's Theory and Methodology*
by V. P. Kuzmin [17]. The author rightly points out that "K. Marx and
F. Engels left us no special study of the systems problem in the form of a
ready-made methodological theory but they did leave a systems theory of
social development and multiple instances of concrete systems solutions of
various questions arising in the study of society as a whole" [17, p. 12].

methods which appeared in the middle of the twentieth century and later became widespread include systems approach, various versions of general systems theory and systems analysis.

Contemporary literature (including Marxist literature) offers rather discrepant definitions of the basic principles and content of the above domains of systems research (see for instance [5], [6], [15], [20], [25], [28], [29], [31]). This is only natural: systems research, especially practical systems studies, are as yet in the making and premature rigid definitions that take no heed of the practical expediency and effectiveness of the theoretical generalizations (still to be obtained and verified) can only hamper the progress of systems research as a whole. Meanwhile we do need some *ad hoc* definitions of systems approach, general systems theory and systems analysis, so in the context of the subject on hand, i.e. the substantiation of materialist dialectics as a philosophic basis for systems research, the following points should be emphasized.

Systems approach, general systems theory and systems analysis are all aspects or sides of contemporary systems research. They have a number of properties in common as well as their own distinctive features. The common properties are:

(1) The theoretical methodological function: each of them is a form of theoretical interpretation of systems research which generalizes the experience of concrete systems studies in its own special terms and serves as a methodological tool for further systems research.

(2) The general, interdisciplinary nature: the objectives of systems approach, general systems theory and systems analysis make it impossible to inscribe any of them in the current subject-wise division of science and technology while the results they yield are applicable to whole complexes of scientific and technical disciplines.

As for the specificity of these three domains of theoretical reflection about systems research, it lies in the peculiar form of reflection inherent in each domain. The specific purpose of systems approach is the formulation of principles, notions and methods of systems research at the level of special methodology with due regard for the general methodology, i.e. materialist dialectics. Systems analysis mostly deals with the elaboration of theoretical methodological procedures for systems studies, the construction of systems and the control of systems involving the human (goal-oriented) factor. The general systems theory, in the opinion of those who were the first to programme such a theory (L. von Bertalanffy [15, pp. 23-82], M. Mesarovic [15, pp. 165-186], K. Boulding [15, pp. 106-124] and others), should be a kind of general science covering all types of systems. However, the practical implementation of the various programmes for the creation of a general systems theory has met with tremendous difficulties. Therefore at the moment general systems theory usually means a more or less generalized version of a systems theory, embracing certain classes or types of systems. Besides, there exists a logical methodological interpretation of general systems theory as a metatheory of systems theories.

Once again I would emphasize the provisional nature of the above definitions which have no claims at all to finality or completeness.* The common and

---

*For a more thorough analysis of the specific features of systems approach and systems analysis see the chapters by V. N. Sadovsky [26] and E. L. Nappelbaum [23] in this book.

specific features of systems approach, general systems theory and systems analysis had to be established for the further elaboration of our thesis: the fundamental role of materialist dialectics at the philosophic basis of systems research.

It follows from the above understanding of systems approach, general systems theory and systems analysis that these spheres of contemporary systems research, though designed to give a theoretical generalization of practical systems studies, are themselves in need of a philosophic interpretation and substantiation. Although systems approach, general systems theory and systems analysis provide theoretical methodological knowledge about systems, none of them claims philosophic generality of global conceptual significance. This role is reserved for materialist dialectics and first of all for the philosophic systems principle, which is a kind of Ariadne's thread for the philosophic interpretation of the role and significance of systems research (see also [7]).

This is how Marx enunciated the theoretical basis of the dialectic approach to systems

> This organic system as an aggregate whole has underlying origins of its own and its evolution toward wholeness consists precisely in subordinating to itself all elements of society or creating therefrom the elements that it still lacks. Thus in the course of historical evolution the system becomes a holistic entity. The development of the system into such an entity constitutes a moment of the system's process and evolution [1, vol. 46, part I, p.229].

The dialectical unity of the systems principle and the principle of development is an important point in the philosophic substantiation of systems research. In this context a system is not just a structured sum of elements but a dynamically organized entity in the process of development. Of course this dialectic unity of system and development cannot but be conflicting. If the contradictory nature of complex systems seems frightening to the metaphysical outlook, the dialectic materialist doctrine regards this very contradiction as the true source of their regular development. It is through rational action on this source that a systems process can be controlled. Allowance for the unity of system and development makes it clear, *inter alia*, that the forecast of a structure cannot be based on a simple extrapolation in time of the present state of the system (e.g. a global system). The parameters of a system and its relation to the environment change in time according to the dialectic laws of progressive advance, hence systematic control can be applied to scientific and technical progress and, more generally, to socio-economic development. In short, the fundamental difference between the dialectical and the metaphysical approach consists in the former's well-founded orientation towards the objective tendency of systems to evolve.

Apart from this ontological side of the matter, there is the methodological side, specifically the role of systems studies in the integration of present day scientific and technical knowledge. Here again materialist dialectics, specifically the philosophic systems principle, is fundamental to the systematic synthesis of knowledge.

The integration of knowledge is an inherently historical problem that has been raised in different forms at various stages of scientific progress. For instance, it has been suggested that the synthesis of knowledge should be achieved on the basis of one leading discipline, and this part of the leader and integrator of all science was assigned to mechanics, physics, biology, or cybernetics. Of course this kind of analysis will go on in spite of its inevitable one-sidedness that makes real integration impossible, albeit it often allows a good grasp of the common features in the methodology of various sciences. But we must stress the importance of systems approach and systems analysis with respect to the integration of knowledge. Within the framework of these methods special sciences keep their independence and qualitative specificity and are not reduced to one leading science, whereas their facts and theoretical structures are centred round the systems techniques which serve as a general method that integrates scientific knowledge so as to increase its efficiency.

The interdisciplinary nature of applied systems studies is essential. It is the realization of an important modern form of knowledge synthesis. Of course this form is as yet insufficiently developed and largely functions as a compensatory mechanism designed not to prevent but merely to abate as far as possible the increasingly harmful effects of the excessive differentiation and occasional isolation of the existing and ever multiplying branches of science.

The comprehensive interdisciplinary systems approach is based on profound qualitative changes in the reality of the scientific, technical and socio-economic development of society, in the means and ends of society control: the objects of control are acquiring giant size and complexity; within each object there is a growing interrelationship of patently heterogeneous factors: economic, social, ecological and technical; there is an increasing interplay of various structural levels; all processes occur at a growing rate; hence the need for a much higher quality and efficiency of control, which should be comprehensive, long-term and goal-oriented [13]. The emergence and development of systems approach is largely due to the objective convergence of the natural and social sciences, the need for a more precise and quantified description of socio-economic processes. All this lays special emphasis on the methodological principles of systems research which stem from the philosophic foundation of materialist dialectics. We now proceed to analyse these principles.

METHODOLOGICAL PRINCIPLES OF SYSTEMS RESEARCH

Sytems research has come into being in response to the growing complexity of an over-technical world, which is largely due to the subject's increased activity seeking to reflect natural laws and transform nature all at the same time. To take one's bearings in this world where the natural interacts with the man-made, one needs new devices which have to be different from those applicable to a purely natural situation. Although the methodology of systems research is still in the making, one already discerns the outlines of some general methodological principles that make systems research distinct from any other kind of scientific endeavour [9, 14].

To demonstrate the specific methodological features of systems research I shall take up systems analysis the way it has been defined above.

(1) An important group of methodological principles underlying systems analysis asserts the organic integrity of the subjective and the objective in systems studies. This group comprises several closely related principles whose main difference lies in the role assigned to the subjective factor in systems research.

The organic integrity of the subjective and the objective is a principle referring to the relationship between research and researcher. The dialectical materialist conception of the subject-object relations in cognition and control stipulates that the difference and interconnection between two types of systems should be taken into account: systems as objects of study and systems as tools for the organization of knowledge, seeking to solve the problems inherent in the object system and important to the researcher.

All special sciences, both natural and social, tackle the first type of systems. Systems of the second type, tools for cognition, control and other forms of human activity, are methodologically the prerogative of philosophy, particularly gnoseology, and of logic and metamathematics. Although systems analysis deals with both types of systems, it owes its name and many of its methodological features to the second interpretation of the term "system". Therefore this term will be more often used in its second sense.

Systems analysis offers a peculiar modification of the traditional subject-object relation of science which requires a rigid enough demarcation line between subject and object.

In fact systems analysis not so much investigates an object *per se* as tackles the problem situation connected with it, seeking eventually to ensure effective interaction with the object. In this sense the objectives of systems analysis are at once narrower and broader than those of traditional research: broader because it requires the additional knowledge of who is to interact with the object and what interactions are therefore possible; narrower, or rather more specific, because our interest is confined to those aspects of the object's behaviour that have some bearing on the effectiveness of the proposed interaction. Naturally the "broadening" and "narrowing" of the object introduce some "extra subjectivity" into systems analysis and bring about a modification of the traditional subject-object relation. Besides, systems analysis normally deals with extremely complex objects, so that to assess different strategies leading to the desired interaction one uses essentially different models of the object. The result is a "target relativity" of the object's description [22, 23].

Furthermore, by virtue of its tendency to subordinate research to the proposed interaction, systems analysis gives a new significance to the principle of the unity of theory and practice which is central to Marxist-Leninist philosophy. Systems analysis consciously renounces the investigation of objects *per se* and conditions all studies by practical requirements so as to attain a pre-set goal. One may say that systems analysis has been called forth by the imperative need to investigate, construct and control those systems that are not as yet described by any mature and reliable theory.

Another aspect of the expanded subjective factor in systems research concerns people who usually enter its objects as important elements and whose individual and collective behaviour eventually determines many essential features of the system's behaviour. If the problem situation depends on the behaviour of various people, then it necessarily depends on the way these people perceive it. Therefore systems analysis lays special emphasis on the adjustment of the external view of the problem situation (on the part of the researcher and "customer") and the internal view (on the part of the system's active element).

Yet another methodological principle of this group, which is closely related
to the previous one, concerns the specificity of the subject-object relation
with regard to "artificial" systems constructed (consciously or unconsciously)
by man. It has been mentioned before that systems analysis deals with large
systems whose integrity is ensured by goal-oriented human activity. This
class includes large technical systems, socio-economic systems and
organizational systems. Each of them, apart from the object side proper,
objectifies the aims, motives, interests, tastes and knowledge of its
designers. One must be able to distinguish between these two components
while analysing the behaviour of such systems. Indeed the first component
determines the real limitations imposed on our interaction with the system
and its control, whereas the second component may, in fact must, be altered
if the earlier aims have lost their attraction or if new knowledge,
technologies or control strategies have emerged.

Speaking of the organic integrity of the objective and the subjective in
systems studies, I should stress the point that subjectivity refers to those
objects which are regarded as manifestations of one and the same system
within the framework of our analysis. In fact the necessity for some kind of
interaction with the object and the resulting problem situation are quite
objective. The same is true of the people's behaviour which predetermines
the system's behaviour. Finally, at any particular time, the aims and
knowledge embodied in the design of a system are also objective. Yet, with
regard to the object placed under study at a given moment in its history or,
more generally, with regard to a whole class of possible interactions, these
components appear in the description of the problem situation as variable,
contingent and subjective.

(2) Another important group of methodological principles underlying systems
analysis is concerned with the notion of structure. One might well say that
the problem of structurization is all but the main distinctive feature of
systems research. It is the structure that gives systems their integrity
and determines the stable characteristics of a system, making it distinct
from other kinds of objects. The notion of structure implies that within
the framework of a given analysis a system will remain unchanged if one of
its subsystems is replaced by another that meets the eventual requirements
for adequate interaction with the other subsystems imposed by the system's
structure. One may say that the integrity of a system is due to the
synthesis of its stability, manifested as its structure, and its relative
variability implied in the generalized description of the system's elements
that only fixes the pattern of their possible interactions within the
structure.

The structures of various systems may be highly diverse. Systems analysis
mostly deals with hierarchic structures. One reason for this is that a
hierarchic structure is so far the most effective way of organizing the data
on an object's behaviour into operational knowledge. Another reason is that
hierarchic structures naturally emerge in the course of development from the
simple to the complex via fixed intermediate forms [27]. Note that the very
notion of a system is hierarchic as it presupposes the unity of three basic
concepts: a system as an element of a larger system which conditions its
interaction with the environment, an integral entity and a sum of its own
elements.

The above observations concerning the role of structure in systems analysis
have far-reaching ramifications for the general strategy of systems research.
Structural analysis is a tool which serves, *inter alia*, to investigate
systems on the basis of functional similarity by solving "the inverse

problem". According to the so-called black box principle, the external behaviour of the various subsystems within a fixed structure is only significant in so far as it conditions their interaction. The internal structure leading to this behaviour is immaterial. On the one hand, this allows a reduction of infinite regression from the system to its subsystems, then to its sub-subsystems, etc. releasing an avalanche-like flow of information relevant to the system's functioning. On the other hand, to approach the system one may first solve "the inverse problem", i.e. project each of the existing subsystems onto a simulated subsystem which would have the required external behaviour though possibly a different structure, and manipulate this model as if it were the real thing.

Functional similarity and simulation are widely used for investigating the important class of adaptive, self-organizing and learning systems. These systems differ from all others in that they are *a priori* plunged into a "superstructure" of interaction with an environment whose elements may change in time, so that the subsystem in question must alter its behaviour to keep the original interaction structure intact.

The structure-conscious approach is closely related to the concept of the organic integrity of the objective and the subjective. The structure of a system should not be regarded as something unique or invariable. Systems analysis rarely deals with unambiguous structures of space distribution; most systems analysed have so-called functional structures or interaction structures whose elements may even be deprived of an absolute space localization. These are not identified *per se* but in the context of the external interaction structure which largely conditions them. It is in this sense that the structuring of a system depends on the problem situation in which the system is considered.

It is therefore possible to distinguish between the so-called goal-seeking (adaptive, learning) systems, as defined in modern cybernetics, and goal-setting systems. In the first case the structuring of the external interaction (a broader system) is imposed from the outside, whereas in the second case it is imposed from within the system itself.

(3) Another important group of methodological principles underlying systems analysis and specifying the dialectic principle of development involves the notion of system dynamics. As a rule the integrity of a system and its structure is only manifest in the process of evolution when a change in one subsystem inevitably entails changes in others. This has been remarked by V. G. Afanasiev who described motion as a condensed expression, "a quintessence of all characteristics inherent in a whole" [5]. It is not only the various subsystems that may change in time; the system's structure also evolves, normally indicating the existence of a higher hierarchic level whose structural stability ensures the system's integrity. The changes may spring from the objective variability of the system *per se* as well as from the variability fo the subject's view, i.e. the variability of the "systems for us". This is an essential point for systems comprising human beings with a changing idea of the system and their part in it.

Of course the dynamic approach is not confined to a mere statement of the system's possible change in time and the importance of studying the change in order to grasp the system's integrity. It also implies a continuous observation of the system's morphogenesis, whose current behaviour can only be understood in a broader context of development. This reasoning may account for the numerous attempts to construct an organismic systems theory which, for all its limitations, justly points out that each system has got its own inertia and its own internal sources of development.

While stipulating that a system's behaviour should be studied on a broader
time scale than the problem situation seems to demand, systems analysis notes
the limited nature of the time frame in which a system can conceivably be
observed and puts forward another methodological principle, that of
observability. According to this principle, feasible observation must enable
the researcher to identify a state of the system which is sufficient for
predicting the system's behaviour in the case of future interactions.

(4) Another group of methodological principles underlying systems analysis
is concerned with its interdisciplinary nature. The basic idea is that
systems analysis deals with extremely complex objects whose description
inevitably involves concepts pertaining to different traditional disciplines,
so that there is a need to adjust various professional languages. The real
meaning of this principle, however, goes much deeper [21]. The point is that
traditional disciplines normally consider the various aspects of a system's
behaviour in deliberate isolation from their cross-relations with other
phenomena which lie within the scope of other disciplines. Systems analysis
renders this isolation impossible, while the interpretation of "traditional"
data in the context of systems interaction is as great a challenge as the
"original" problem itself. This is another manifestation of the well-known
systems effect: once a collection of facts forms a system, it acquires a new
quality and is no longer tantamount to a simple sum of the initial facts.

Besides, the interdisciplinary approach insists on the coordination of the
highly specialized paradigms that underlie (often not manifestly) the theories
of individual phenomena relevant to a system's behaviour. *Inter alia*, this
principle justifies the necessity for a correct combination of the formal and
the informal components of systems research (see [18].

(5) The organic unity of the formalized and the nonformalized is a major
methodological principle of systems analysis. The basic advantages of
systems analysis are often made out to be its high degree of formality and
the application of numerical methods to problems that used to defy
formalization. I believe that the main achievement of systems analysis lies
elsewhere. It has already been pointed out that, for the first time ever in
specialized scientific knowledge, systems analysis overtly acknowledges the
major role of the "subjective" elements in control problems and elaborates
scientific methods allowing for these elements.

Indeed, success primarily depends on whether one has correctly set the goal
and defined the problem, and only secondarily on how well the problem has
been solved, which is where formalization comes in. Hence it is emphatically
recommended that the researcher should start with a thorough and comprehensive
study of the problem, which is never a thing in itself but a reflection of
the opinions and concerns of a certain group of people. The very first thing
to do is to find out what they want and why. The more natural and
"objective" the problem looks, the more closely and meticulously it should be
investigated. The formulation of a problem does not always strictly
correspond to its essence. Only too often an imperceptible and unconscious
substitution is made at this point, resulting in an arbitrary imposition of
unfounded limitations on perfectly admissible ways of solving the problem.
Therefore systems analysis requires that the researcher should understand
and make extensive use of the dialectic relationship between the formulation
of a problem and the choice of methods for solving it.

Not a single alternative solution should be rejected *a priori*, on the basis
of "self-evident" considerations, without a thorough analysis (including an
investigation of the reasons why it should not be considered). As a rule it
is not a search for the best of the known solutions but the discovery of some

utterly new "dimension" of the space of alternatives that yields truly effective solutions. Therefore it is as important to look for new alternatives as to look for the best of the available ones.

Speaking of modelling in systems analysis, I do not imply that all models should be quantitative. A model may be qualitative or verbal, but it should by all means identify all the major premises and hypotheses on which it is based; these should, as far as possible, refer to the most elementary level of the model's organization and be absolutely clear and unambiguous. Then it would be possible to assess the situation's sensitivity to the assumptions made and to find out which of them are crucial and should therefore be most closely tested. The above refers to all models, including mathematical ones. It is very important to be able to quantitatively describe the object's behaviour, but we must not forget that a model is effective in so far as its basic premises are correct.

The dialectic unity of formalized and nonformalized treatment is a characteristic feature of systems approach and systems analysis. Two points here are significant. First, it is erroneous to believe that the model's formalized components alone enter systems research while the nonformalized ones are left out; on the contrary, systems studies should include the nonformalized premises. Secondly, it is not correct to identify the formalizable and the quantifiable, as new and more subtle formalization techniques are increasingly coming into use. The dialectic unity of the formalized and the unformalized, of the quantified and the unquantified, is crucial to the practical realization of systems research.

To achieve the unity of theory and practice in systems analysis it is essential to supply the model with a correct selection of criteria. Therefore the main quantitative component of systems analysis consists in a broad use of different criteria, or rather performance indices, for the various alternatives of the problem's solution. Systems analysis points out that no performance index can adequately characterize the manifold assessments of the alternative in question, hence there is an imperative need for vectorial quality criteria which should contain various estimations of the solution's efficiency as well as evaluations of its costs, including the costs of research and implementation. Any performance index is implicitly based on some model of the situation and cannot be understood correctly without an indication of the model's nature.

Once the need for vectorial criteria was realized, inevitable changes occurred in the search for solutions. If until recently the basic principle has been that of optimization, i.e. the search for an alternative which would best meet one scalar criterion, at present the multicriterial approach is winning recognition.

The fact that systems analysis is practice-oriented by no means implies a deficiency in respect of theory. On the contrary, this orientation reflects the changed role and nature of scientific theory at the time of scientific and technical revolution. Systems analysis and systems research as a whole, while keeping within the current of the age-old tradition of rational cognition, answer entirely new questions raised by the modern epoch.

REFERENCES

1. K. Marx and F. Engels, *Collected Works*, 2nd edition, Moscow, in Russian.
2. V. I. Lenin, *Collected Works*, Vol. 14. Moscow, Progress Publishers (1972).
3. L. I. Brezhnev, *Report of the CPSU Central Committee to XXVth Congress of the CPSU*. Moscow, Novosti Press Agency Publishing House (1976).
4. V. G. Afanasyev, *The problem of integrity in philosophy and biology*. Moscow (1964), in Russian.
5. V. G. Afanasyev, On the Systems Approach to Social Cognition, *Voprosy Filosofii*, 6, 98-111 (1973), in Russian.
6. I. V. Blauberg and E. G. Yudin, *The development and essence of the systems approach*. Moscow, Nauka (1973), in Russian.
7. I. V. Blauberg, V. N. Sadovsky and B. G. Yudin, The Systems Principle in Philosophy and Systems Approach, *Voprosy Filosofii* 8, 39-52 (1978), in Russian.
8. D. M. Gvishiani, *Organization and management*. Moscow (1972), in Russian.
9. D. M. Gvishiani, Dialectics and Principles of Systems Analysis, *Systems analysis and management of scientific and technological development: Abstracts of Papers, Theoretical Conference, Moscow* p. 3 Obninsk (1978), in Russian.
10. D. M. Gvishiani, Methodological Problems of Global Development Modelling, *Voprosy Filosofii* 2, 14-28 (1978), in Russian.
11. G. V. F. Hegel, *Encyclopaedia of Philosophic Sciences*, Vol. 1, *Science of Logic*. Moscow, Mysl' (1975), in Russian.
12. V. A. Ghelovani, A Modelling System for Global Development Processes, in: *Methodology of Systems Analysis*, Moscow, Proceedings of the All-Union Institute of Systems Studies, Part I, No. 6, 73-82 (1978), in Russian.
13. V. I. Danilov-Danilyan, Goal-Oriented Programmes and Optimal Long-Range Planning *Economika i Mat. Metody* 3, 1150-1163 (1977), in Russian.
14. S. V. Emelyanov and E. L. Nappelbaum, Fundamental principles of systems analysis, in: *Problems of Management in Socialist Industry*. Moscow, Economika (1974), in Russian.
15. *Studies in General Systems Theory*, Collection of translated papers (V. N. Sadovsky and E. G. Yudin Eds). Moscow, Progress Publishers (1969), in Russian.
16. E. S. Quade, *Systems Analysis and Policy Planning: Applications in Defence*. New York, Elsevier (1918).
17. V. P. Kuzmin, *The systems principle in Marx's theory and methodology*. Moscow (1976), in Russian.
18. N. I. Lapin, Typology of nonformalized components of a modelling system, in: *Methodology of systems analysis*, Moscow, Proceeding of the All-Union Institute of Systems Studies, Part I, No. 6, 5-13 (1978), in Russian.
19. O. I. Larichev, Systems analysis: problems and trends, *Automation and remote control* 2 (1975), in Russian.
20. V. S. Levchenkov and A. I. Propoy, On the general systems theory, Moscow, All-Union Institute of Systems Studies, preprint (1978), in Russian.
21. E. M. Mirsky, Interdisciplinary studies as an object of scientological investigation, in: *Systems Research Yearbook*. Moscow, Nauka (1972), in Russian.
22. E. L. Nappelbaum, Some aspects of the problem of uncertainty in systems analysis problems, in: *Systems analysis and management of scientific and technological development: Abstracts of Papers, Theoretical Conference, Moscow*, pp. 46-50. Obninsk (1978), in Russian.
23. E. L. Nappelbaum, Systems analysis as a research programme - guidelines and objectives, this book.

24. I. B. Novik, *On Modelling of Complex Systems*. Moscow (1965), in Russian.
25. V. N. Sadovsky, *Fundamentals of the General Systems Theory: Logico-methodological Analysis*. Moscow, Nauka (1974), in Russian.
26. V. N. Sadovsky, The systems approach and general systems theory – state of the art, main problems and prospects for development, this book.
27. H. Simon, *The Sciences of the Artificial*. Cambridge, Mass., MIT Press (1969).
28. V. S. Tukhtin, *Reflection, Systems, Cybernatics*. Moscow (1972), in Russian.
29. A. I. Uemov, *The Systems Approach and General Systems Theory*. Moscow (1978), in Russian.
30. J. G. Wirt, A. J. Lieberman and R. E. Levien, *R & D Management*. Lexington, Mass., Lexington Books (1975).
31. Yu. A. Urmantsev, *The Symmetry of Nature and the Nature of Symmetry*. Moscow (1974), in Russian.
32. P. N. Fedoseyev, Philosophy and integration of knowledge, *Voprosy Filosofii* 7, 16-30 (1978), in Russian.
33. E. G. Yudin, *The Systems Approach and the Principle of Activity: Methodological Problems of Modern Science*, p. 391. Moscow, Nauka (1978), in Russian.

SYSTEMS APPROACH AND GENERAL SYSTEMS THEORY — STATE OF THE ART, MAIN PROBLEMS AND PROSPECTS FOR DEVELOPMENT

V. N. Sadovsky

INTRODUCTION

Modern systems studies embrace a broad set of scientific, technological, and organizational-managerial problems and the specific methods of their solution. There is a need to theoretically justify systems studies as a rapidly developing sphere of science and technology in the second half of the twentieth century and as a special form of scientific and technological activity. This justification involves the solution of two interrelated problems: the definition of the philosophical foundation of systems studies and the formulation of intrascientific methodological systems concepts. Once the first problem is solved it is possible to establish the Weltanschauung significance of systems research and describe its fundamental philosophical prerequisites. The solution of the second problem requires the intra-scientific theoretical justification of systems studies, which should result in the construction of a methodological basis for various specific systems developments.

In the late 1950s the methodological studies of systems research entered a period of intense growth, and in the course of its nearly 25 year-history this sphere of modern methodology of scientific and technological knowledge has become quite successful, both from the quantitative (the range of the problems being studied) and the qualitative (the results obtained) point of view. This statement can be confirmed by the conclusions of two recently conducted scientometric analyses of both Soviet and Western literature (primarily in the English language) on methodological problems of systems studies. The first, by S. I. Doroshenko [15], dealt with the Soviet literature on the subject published during 1957-1974. The quantitative indicators of this set of publications - 1074 papers and 688 authors - are quite sufficient to establish that there has been a very rapid development of research in this sphere of knowledge. This conclusion can be made both more precise and specific if the "internal" scientometric indicators of this set of publications are determined.

Between 1957 and 1974 the average overall increase of the set was 20% annually, the figures for papers being 24% and for books 17%. An almost exponential growth can be seen. The increase was especially high from 1970 to 1974; during this period 679 units were added to the set, whereas the preceding 13 years saw an increase of only 378 publications. In 1971-1974 the set was doubled, and although in the earlier period (1957-1970) the doubling of the total number of publications required a somewhat greater period of time, the average doubling indices of the Soviet literature publication set on methodological aspects of systems research are considerably higher than the similar indices even for rapidly developing branches of science with a doubling period of 7-12 years. The growth of the number of authors in this publication set is also substantially higher than average, especially in 1971-1974, when the annual increase in the number of new authors exceeded 30% (for more details see [15]).

Another scientometric study devoted mainly to systems-methodological literature published in 1927-1976 in English [48] produced similar results. The total volume in this set came to 1409 sources, while its growth rate is very close to that of the corresponding Soviet literature publication set.

From this scientometric data it is possible to say that the methodology of systems studies is developing extremely rapidly not only as to the growth rate of the number of publications, but also as to the rate of new authors entering this sphere.

The reasons for this interest in the methodology of systems research are quite obvious, and have been frequently discussed in the literature (see for instance [6]). The main factor is that in the second half of the twentieth century it has been realized that in practically any branch of his activity (science, technology, industry, management, etc.) man deals with complex interrelated objects representing various kinds of systems, rather than with individual isolated units, things, processes or phenomena. And the task of studying (designing, managing, etc.) systems inevitably led to the problem of the methodology of such studies (design, management). Moreover, it is only when a certain level in the formulation of methodological problems had been achieved, that one could expect to be successful in the specific analysis of systems. This was exactly the reason for the truly avalanche-like flow of systems-methodological studies and publications that appeared during the last two decades.

It is certainly not possible to consider in a single paper the entire range or problems of systems research, i.e. those that include both philosophical and specialized-methodological points of view, and this is not the objective. My purpose is to use the experience of development of theoretical and methodological aspects of systems research to identify the specificity of the intrascientific forms of reflection concerning the methods and principles of systems analysis, primarily the systems approach and the general systems theory, to try and evaluate the results achieved in these fields, and to outline their evolutionary perspectives. One other limitation of the objective of the present paper relates to its historical framework. Although the systems interpretation of the objects of knowledge and technological design originated centuries ago (see in this connection [22], [30],[55]), I will be concerned in this case only with the present stage of systems studies, which was initiated in the first post-war decade.

SPECIALIZED SYSTEMS STUDIES AND THE METHODOLOGY OF SYSTEMS RESEARCH

The very first publications specifically devoted to systems problems appearing at the end of the 1940s and 1950s (see [20], [24], [43], [50, Chapters 2 and 3]) stressed that the field of systems studies covered not only the development of philosophical and methodological principles of systems analysis, but also specific systems research in various spheres of science and technological design. This statement reflected the real tendency of the evolution of systems problems, which very soon led to the formation of a special scientific and technological area — systems studies in the broad sense. This includes the whole set of modern scientific, technological and industrial-practical systems, problems and concepts. Systems research in the broad sense appears as a certain aspect of the whole range of human activity, since systems representation can be constructed for any of its objects (and frequently the practical expediency of this kind of "seeing" the objects has been proved).

The application of the principles of systems theory to systems studies themselves required that the analysis of the *system* of system studies should also be specified. And as far as this factor is concerned, the identification (and the establishment of interrelations) of specialized systems studies on the one hand, and the methodology of systems research and analysis on the other, is of basic significance. Although this conclusion seems to be an obvious consequence of the relation between methodological and specialized scientific knowledge that is traditionally considered in philosophy, in this case it implies the determination of the principal directions of systems research development and the establishment of conditions under which these trends can be realized.

Where the sphere of systems studies as a whole is divided into specialized systems studies and the methodology of systems analysis, the stress is laid firstly on the basic difference between the problems solved in each of these fields (in the former instance, it is the construction of models of specific systems and examination of the results; in the latter it is the formulation of methodological principles that can aid in the design and investigation of such models); second the difference between the means of the analysis (the analytical tools of specialized research, in the former, and the methods of conceptual (philosphical and methodological) analysis of scientific knowledge combined with the formal apparatus of modern logic and some branches of mathematics in the latter); and third the difference between scientific communities working in the first and second fields (the scientific community dealing with specialized systems studies consists of people representing specific scientific and technological disciplines who are oriented towards systems "vision" of the objects being studied or designed; the scientific community of systems research methodologists includes some of those methodologists and logicians of science who specialize in modern epistemology).

For all the peculiarities of specialized systems research and the methodology of systems analysis mentioned above, there are obvious connections between these spheres, in the sense that the second should supply the specific means for the systems analysis of the first; no productive development of the second sphere is possible without the specific material obtained in the first sphere, etc., and even certain intersections in the set of problems and composition of the corresponding scientific communities. But despite the presence of these common features, which, incidentally, is exactly the reason for combining these two different fields into a single sphere of modern systems research, the realization of the difference between them is not only fundamental, but also necessary for the progress of both specialized systems studies and the methodology of systems research.

THE SYSTEMS APPROACH

To denote the set of methodological problems of systems studies, the literature, especially Soviet philosophical literature, usually employs the term "systems approach" (see for instance [5]). Considerable success has recently been achieved in the definition of specific features of systems approach as a certain methodological concept [7]. This resulted in the following basic conclusions.

In modern science and technology, the systems approach appears as a special methodological concept called upon to formulate in a systematic manner the totality of methods for the analysis and design of systems belonging to various types and classes. It is important to emphasize that this interpretation of the systems approach does not imply its ability to solve problems of general, philosophical methodology (this being the "internal" problem of philosophy), although the methodological knowledge obtained in it may specify and further develop the corresponding branches of philosophical methodology.

To understand the peculiar features of the systems approach it is necessary to adopt the historical point of view which regards the systems approach as a certain stage in the evolution of epistemological methods, the methods of research and design activity and the techniques for describing the nature of objects under study or those created artificially. As far as its history is concerned the systems approach is replacing the methodological concepts of mechanicism and elementarism, as its specificity and tasks are opposed to those of these theories. Thus, to determine the place and functions of the systems approach in scientific knowledge, the relation "mechanism-system" is basic; it establishes the difference and interaction between various ways (mechanistic and integrative) of representing objects, as well as between various methods of analysing them (on the one hand, elementarist and reductionist, and on the other synthetic, anti-reductionist, integrative). Accordingly, the systems approach is primarily intended to be used for the development of techniques for the analysis and synthesis of objects, the description of their integrative characteristics resulting, in particular, from representation of studied and designed objects as goal-oriented systems; the synthesis of "element" and "integrative knowledge" of the above objects; the investigation of the relation between given and other systems which constitute their environment, etc.

Like any other methodological concept, the systems approach "addresses itself", first, to the corresponding specialized scientific and technological (systems) knowledge and second, to the methodology of science as a whole, moreover, to philosophical methodology. These two relations are essentially different. As far as the scientists working in specific scientific and technological fields are concerned, the systems approach assumes the form of systematization of systems research methods and principles that can be used by experts in their work. The problem of employing the systems approach in specialized scientific and technological analysis is primarily a problem for the experts concerned. Besides, the systems approach which can be regarded in each given period of time as a certain combination of methodological statements, requires development and perfection. And the elaboration of the systems approach can occur only in the general context of specialized-methodological and philosophical-methodological analysis, the solution of this problem falling to the corresponding groups of philosophers and methodologists of science. This development, which was extensively reflected in the systems literature of the past two decades, was accompanied, on the one hand, by making the relations between philosophy and the systems approach more precise,* and on the other, by identification and study of various forms of theoretical reflection concerning systems research, i.e. the general systems theory, logic and methodology of systems studies, etc.

LEVELS OF CONTEMPORARY METHODOLOGICAL KNOWLEDGE. THE PHILOSOPHICAL SYSTEMS PRINCIPLE AND THE SYSTEMS APPROACH

The concept of various levels of contemporary methodological knowledge formulated at the end of the 1960s and the outset of the 1970s by several Soviet philosophers ([7, pp. 65-84; 25-33, Chapter 1]) played an important role in revealing the essence of the systems approach and in determining its links with the philosophy of dialectical materialism. This concept has been brought to life by an intense process consisting in the differentiation of methodological knowledge which has occurred in the twentieth century alongside the rapid development of science, the increasing complexity of its structure, the substantial increase of the application of theoretical, abstract reasoning, the broad use of mathematization and formalization, etc. It would be appropriate to single out four fundamental levels.

(1) The level of *philosophical methodology*, i.e. the analysis of general principles and the category structure of science on the whole. This area of methodology represents a branch of philosophical knowledge and is being elaborated by methods specific to philosophy.

(2) The level of *general scientific methodological principles and forms of research*, which includes informal general scientific concept (the methodology of cybernetics, and the methodological foundations of interdisciplinary studies, etc.) as well as formal methodological theories (the theory of the hypothetical-deductive structure of science, inductivist methodology of scientific knowledge, etc.). As a matter of fact, the significance of the general scientific property of the second-level methodological concepts is that their nature is interdisciplinary. Since general scientific methodological concepts do not claim the ability to solve the general philosophical problems and those of Weltanschauung, they are being developed primarily in the framework of modern logic and the methodology of science.

(3) The level of *specific-scientific methodology* corresponding to the analysis of the methods, principles and procedures used in specialized scientific disciplines. The central objective of this methodological level is the statement and description of the totality of methodological techniques and principles specific to a certain field, e.g. physics, biology, chemistry, psychology, sociology, etc. The prime difference between the second methodological level and the specific-scientific methodological concepts is that the latter are "tied in with the object", i.e. their statements and

---

*The result of the discussion of these problems in Western philosophical literature was the formulation of the concept of systems philosophy [53], [55]. It should be noted that this concept did not gain broad acceptance even among Western systems theoreticians, principally because of its extremely speculative nature and the confusion between the forms of specificity of scientific and philosophical knowledge (for details see [8]).

recommendations pertain to a strictly limited class of objects and knowledge situations specific to a given scientific field. It is the unified efforts by the methodologists of science and the theoreticians of the corresponding spheres of knowledge that are responsible for the construction of specific scientific methodological concepts.

(4) The level of *research methods and techniques* - the description of ways of obtaining relevant information, the experimental conditions, taking into account errors, methods of experimental data processing, etc. At this level methodological knowledge meets methodological requirements and regulations. The latter are usually closely linked to the specificity of a certain scientific or technological discipline, and are elaborated in the framework of specific scientific knowledge.

From the point of view of this classification, the systems approach belongs to the second level of methodological knowledge. A detailed justification of this statement can be found in [7]; the result was the systems approach assessed both negatively (as a nonphilosophic methodology), and positively (as general scientific, interdisciplinary methodological knowledge). According to the concept of methodological levels, philosophic methodology (dialectics) is basic to any form and type of methodological knowledge, and each more general methodological level conditions the levels of methodology dealing with less generalized statements. This served as the foundation for the natural explanation of both the formula "dialectics - systems approach" (the connection between philosophical methodology and general scientific nonphilosophic methodological concepts) and the relative independence of the systems approach, combined with its role in the formulation of specific scientific methodological and systematic principles and statements.

It would appear that the above interpretation of the status of the systems approach, as well as its interaction with dialectics, adequately takes into account the modern level of theoretical realization of the essence and specificity characterizing the methods of systems studies and the assertion, fundamental to Marxist philosophy, of the primacy of philosophical methodology with respect to other levels of methodological knowledge. This interpretation has been quite broadly reflected in Soviet methodological literature, particularly in many articles published in the *Systems Research Yearbook* [36].

At the same time, recent Soviet systems literature additionally features a different treatment of the systems approach. In this sense the most significant is [29] regarding the systems property as "the most important feature of the dialectical method" [29, p. 96]. Accordingly, in his analysis of the role of the systems approach in Hegel's dialectics, the author concluded that "the entire totality of categories in Hegelian logic is nothing but the category structure of the systems approach" [29, p. 98]. Thus in this interpretation, the systems approach appears as an essential feature or aspect of dialectics.

It appears as though one is dealing with two fundamentally different concepts of the systems approach and in this connection it is necessary to clarify to what extent each is valid. I think, however, that the real difference between these two concepts is terminological rather than that in meaning. To demonstrate this, the philosophical systems principle must be distinguished from the systems approach. The first "means that a phenomenon of objective reality regarded from the point of view of all the regularities of the system and the interaction of its components forms a special epistemological prism, or a special 'dimension' of reality" [22, p. 10]. As can be shown (and this was convincingly done in [22, 29, 30]) the systems principle interpreted in this way and embracing the philosophical ideas of the systems connection between the elements of objects, wholeness, structural property, etc. was important in the history of the development of all philosophy and was best expressed in the Marxist-Leninist philosophical conception. One would be therefore completely justified to say that *"Marxist theory and methodology contain the systems principle as one of their most essential components"* [22, p. 20]. I would like to emphasize the fact that the point in question here is exactly the philosophical systems principle and not the systems approach. As for the latter, its treatment as the general scientific nonphilosophical methodological concept, presented specifically in this paper, seems to be the most adequate, both from the point of view of history (the term "systems approach" itself was introduced literally several years ago) and of meaning (the systems approach principles do not claim philosophical generality). I would go on to say further that the notion of "approach" (in the sense of a research approach) is nonphilosophical; philosophy usually deals with "methods", "fundamental principles of research", etc., rather than with "approaches". This is another reason why the term "systems approach" should not be transferred, with its content radically altered, to the sphere of philosophical methodology, which has never used this concept and can employ other quite adequate terms, such as the "systems principle" and the "systems method" for denoting the corresponding field.

Thus, the above terminological difference suggests that the philosophical ideas of systems property, wholeness, structural property and the universal nature of connections between the objects of reality that appear in the form of the systems principle - should represent an important aspect of philosophical methodology and an essential facet of the dialectical method. For the construction of all other forms of theoretical reflection (both general and specialized scientific) with respect to systems studies, including the systems approach, the systems principle is of fundamental methodological significance. And the main task of the systems approach, which does not claim the ability to solve the philosophical problem of systems research, is the development of general scientific, interdisciplinary scientific concepts and methods and ways of analysing systems objects. This understanding of the essence of the systems approach and its relation to dialectics can be considered one of the results of the evolution of systems research methodology in the USSR (see also [9, 13]).

THE GENERAL SYSTEMS THEORY

It was previously noted that the systems studies that have been recently conducted with great vigour paid considerable attention not only to the elucidation of the interaction between philosophical methodology and the systems approach, but also to the analysis of various forms of scientific and theoretical reflection concerning systems research, first of all, the question of the status and the methods of construction of the general systems theory. A very detailed treatment of these issues can be found in [1, 3, 7, 17, 21, 27, 32, 33, 35, 40-42, 44, 46, 50, 52-54].

Despite the diversity of the methods of constructing the general systems theory currently suggested, two main directions, or two basic trends, in the development of this field can be identified. The first direction interprets the general systems theory as a theoretical construction describing all possible kinds and types of systems, or, to put it in another way, as the *object* general systems theory [41, 42]. According to one of the advocates of this approach, in the final analysis the general systems theory should "give the researcher some kind of a list of (1) what *must* be, (2) what *can* be, (3) what *cannot* be for systems, material and/or ideal" [42, p. 51]. The second direction treats the problems of the general systems theory primarily as methodological, associated with the theoretical description of the methods of system analysis and the methods of constructing various specialized systems theories. In the framework of this approach, the general systems theory appears as a *systems metatheory*, i.e. as the general theory of systems theories [33, pp. 51-76, 195-203]. It would be appropriate to identify and compare the initial assumptions made by the proponents of the above two trends in the construction of the general systems theory.

It is a well-known fact that any scientific theory describing a certain class of objects is general and universal in the sense that if this theory satisfies the norms of theoretical construction adopted by a scientific community it contains the knowledge of any object of this class. A "general" theory understood in this way would include any specifically formulated version of the general systems theory (i.e. those suggested earlier by A. A. Bogdanov, L. von Bertalanffy and others [10, 17, 50], as well as more recent constructions of this kind now being developed by G. J. Klir, M. Mesarovič, A. I. Uemov, Yu. A. Urmantsev and others [27, 41, 42, 52]). Because of the above statement however, the adjective "general" as applied to any version of the systems theory does not furnish any additional generality to such theoretical structures. Any given theory is general for a given class of objects, its generality consisting in the fact that it is a theory if it is able to solve the theoretical problems facing it. Therefore, strictly speaking, any additional description of such concepts corresponding to the term "general" is either superfluous or has a different meaning.

Indeed, systems studies theoreticians presenting various projects for constructing the object general systems theory underline that there is an essential difference in the degree of generality of different variants of the general systems theory. In their opinion, alongside specialized systems theories such as the theory of biological systems, the theory of psychological systems, the theory of automata, etc. there are more general systems theories, like the theories of cybernetics or the theories of management systems, and the ideal thing would be to construct a truly general systems theory which would combine all more specialized systems concepts and formulate the systems principles applicable to all possible systems. Certainly, such a theory would be universal in the sense described above (I shall call it the "first sense of the universality of a scientific theory"); in addition, according to the theoreticians alluded to earlier, it should also be universal in a second sense, i.e. it should contain all fundamental information concerning the general properties, relations and connections between all existing and all possible systems. Thus, if in the expression "general systems theory" the term "general" has a meaning of achieving the universality of a theory in the second sense, the expression "general systems theory" itself is quite justified and correct; but the question arises as to whether such a theory is possible in principle.

Of course, it cannot be doubted that different systems theories are characterized by different degrees of generality. It is indisputable, for instance, that, compared with the general system theory of L. von Bertalanffy [17, pp. 23-82], which in fact happens to be simply a certain generalization of the theory of open systems (i.e. it is, at best, a theory of some classes of material systems), not only the mathematical general systems theory of M. Mesarovič [17, pp. 165-180; 27], but also the parametric version of the general systems theory developed by A. I. Uemov [41], as well as other similar constructions, are more general and include far broader classes of systems objects. But is it possible to suggest that such theories can in principle contain all the fundamental information concerning systems, or in other words, that they are general and universal in the second sense?

It would appear that the answer to the question should be negative. The two arguments below support this conclusion.

(1) The measure of generality of any version of the systems theory is determined by the degree of generality of its basic concepts, primarily the main concept of such a theory, the concept of system. The following difficulty arises in the specification of the extent of the generality of the concept of system. The choice of the most general concept of system, which might include all the objects of the world and of knowledge obviously poses the threat that the content of the theory based on this concept would be trivial, while because of a certain reasonable limitation of the generality of the concept of system, a theory designed on this basis turns out to be a systems theory only for certain classes of systems objects, but not for all possible systems. With this difficulty taken into account, most systems theoreticians deliberately impose constraints on the class of objects that fit into their definition of the concept of system (according to M. Mesarovič, the system is a relation of a family of sets; other authors define the system as some class of mathematical models, etc.), they use it as a basis for constructing the theory of the corresponding types of systems and, if this theory is described as the general systems theory, it is done rather because of tradition. At any rate, it is not implied that their theory is universal in the second sense.

As far as I know the only attempt to overcome this difficulty (i.e. conscious introduction of the most general concept of system and the conviction that a general systems theory for non-trivial objects can be constructed on this basis) was undertaken by A. I. Uemov [40, 41]. But it seems that his parametric general systems theory cannot claim universality in the second sense. The point is that for all its sufficiently high degree of generality Uemov's concept of system (according to his definition, the system represents any object possessing certain properties that form some relation specified beforehand [41, p. 121]), coincides extensionally with Mesarovič's formulation of the system (in which the latter is understood as a relation defined on a family of sets [27, p. 167]) and therefore is as limited as this formulation. Besides, the following argument (2) can also be used against the understanding of the parametric systems theory as the *general* object theory of systems.

(2) It is obvious that systems are not simply material formations of some kind, but various abstract constructions as well. In particular, scientific theories, logical and mathematical calculi, relatively independent stages of research such as empirical studies, and finally, the methodology of science as a whole, as well as its individual branches (for instance, the methodology of systems research) can be called systems. Therefore, the requirement that the general systems theory should formulate all fundamental information concerning all possible systems implies that such a theory should provide generalized knowledge of not only all kinds and types of material systems, but also of ideal, conceptual systems, including the methodology of systems research. Understood in this way, the general systems theory should cross the boundaries of specialized-scientific (object) knowledge. The set of its problems will then at least overlap that of the philosophical analysis of systems. No positive results can be obtained from

syncretic unification of problems in specialized-scientific and philosophical studies either from the practical or the theoretical point of view, as has been demonstrated convincingly by the entire history of the development of human knowledge. It is exactly this which is the most important argument against the attempts to construct a general object systems theory which claims to be universal in the second sense.

Realization that it is fundamentally impossible to formulate a general object systems theory is the starting point of the second trend in the development of the general systems theory alluded to previously. A consequence of the recognition of this fact is that no matter how the problems of the general systems theory are interpreted the requirement that this theory should be universal in the second sense should be discarded. There are two possibilities in this situation: (a) within the framework of the object analysis of systems, the systems theories that are more general than clearly limited specialized system-theoretical constructions are regarded as different versions of the general systems theory, and (b) the grounds for the generality of a systems theory are looked for not in the object but in the methodological and metatheoretical aspects of the analysis. In the first case, I in fact remain within the limits of the specialized systems theory (this has already been discussed); in the second case - although also without claiming that universality can be achieved in the second sense - I would be more justified to assume that the resulting system-theoretical construction would correspond at least to the intuitive meaning of the concept "general systems theory".

Indeed, if the objective of a general systems theory is in fact the construction of a systems metatheory, it should at least in principle formulate the knowledge about the methods of construction (the corresponding languages, the logical apparatus, the methodological procedures, etc.) of all possible systems theories and for this reason its degree of generality is obviously higher than that of various object systems theories. In a systems metatheory, the second level, i.e. not only the knowledge of specific (real as well as abstract) systems but the principles of design of systems theories as well, is generalized. The result is that although such a theory certainly does not contain all fundamental information for all possible systems - its limitations being quite obvious - it is still far more able to perform the functions of a general theory of systems. Therefore, this interpretation of the status of such a theory has recently been used as the basis for several attempts at constructing a systems metatheory (see for instance [33, 51, 52]).

The discussion of the nature and status of the general systems theory which has recently been quite important in systems literature and is of course described here only from the point of view of its most general aspects is not just of purely theoretical interest. The result of this discussion is that now the specificity and the interrelation between various forms of scientific and theoretical reflection concerning systems studies, the functions which the systems approach is expected to perform in these studies, the specialized systems theories and the general theory of systems are understood far more profoundly than at the end of the 1950s. Undoubtedly this theoretical knowledge exercises its influence on the direction and the methods of the study and design of specific systems in various fields of modern science, technology and practical activity. Therefore, the results of this discussion can be regarded as a significant outcome of systems studies development in the period between the 1950s and the 1970s.

SYSTEMS ANALYSIS

Without at least a brief outline of the problems of systems analysis the above picture of the main trends in the development of a methodology of systems research would be largely incomplete. This direction of system studies is apparently the youngest, but it is the point around which the principal efforts of systems theoreticians and practical scientists have concentrated in the past 15 years. This is primarily due to the explicit practical orientation of systems analysis.

As far as its history is concerned, systems analysis appeared as the next stage in the evolution of operations research and systems technology which was so sensationally successful in the 1950s and 60s. Like its predecessors, systems analysis (or the analysis of systems) represents, first of all, a certain type of scientific and technological activity necessary for the investigation and design of complex and supercomplex objects. The fact that it is theoretically and practically impossible to find analytical solutions to problems such as environmental pollution, sufficient food supply for the world's population, the construction of global models of development, etc., means that these problems are regarded as certain complex systems whose analysis demands the entire arsenal of existing research methods, including various heuristic techniques. Understood in this way, systems analysis is a special type of scientific and technological art; it leads to significant results if used by an experienced professional and becomes practically useless when applied mechanically or unimaginatively (unfortunately, the latter situation is at least as frequent as the former).

As with any art, to be successful, i.e. to satisfy the objective functions formulated beforehand, systems analysis should first be based on a certain theoretical foundation, and second, create in the process of its application models for future use. It is generally recognized in the literature that the theoretical foundation of systems analysis is provided by the principles of the systems approach and the general systems theory (see [16, 18, 28]). But in this case, these principles, instead of appearing in what could be called their theoretical purity, are associated with a definite class of systems: social, economic, man-machine, etc. - in other words, with the main object of systems-analytic activity. Recently theoretical substantiation of this activity led to the emergence of several new directions in systems studies: systems dynamics, heuristic programming, simulation, whose totality constitutes what one would call theoretical systems analysis to distinguish it from applied.

A by-product of this theoretization of systems analysis, which is beyond doubt progressive and extremely important, is the point of view expressed rather frequently in the last few years which amounts to a statement that systems analysis covers the entire theoretical and methodological field of systems studies (see for instance [49]). Usually researchers do not argue over terms, if these are defined sufficiently rigorously and it is quite likely that systems specialists will eventually agree that systems analysis is a synonym for systems research. But it would be incorrect to discard hastily "operating" scientific terms, especially if they convey a clear meaning (and it seems that the concept "systems analysis" does). The situation discussed here is very similar to the concept of "systems approach" which was dealt with previously.

As far as both history and content are concerned, systems analysis, as well as systems approach, is characterized by a quite definite meaning, namely, that of a combination of methods and techniques for making decisions in the design, construction and management of complex and supercomplex objects (social, economic, technological, etc.). When compared with the general methodology of systems research, systems analysis exhibits two kinds of limitations: one is postulated by the type of objects under consideration (it is interested only in artificial (man-made) objects), and furthermore, the human being plays an extremely important, if not decisive, role in their activity; the other is postulated by the nature of the systems problems under study (associated mainly with decision-making and management). Hence the high degree of attention that systems analysis gives to goal-oriented functioning of systems (see [2, 28]). To put it in another way, as far as systems analysis is concerned the system is goal-oriented which naturally does not exhaust the entire class of systems subject to scientific study.

Based on the above the difference between the functions and problems of systems analysis and systems approach becomes quite obvious. Not only systems analysis and systems approach, but also the general systems theory represents in its theoretical and methodological aspects, various forms of the development of systems research methodology and the attitude of systems analysis to the systems approach is that of a part in respect to the whole. Systems analysis considered from the practical point of view is the application of systems principles in the investigation of the decision-making process and management of complex social, economic and technological-engineering systems.

UNRESOLVED QUESTIONS IN THE DEVELOPMENT OF METHODOLOGY OF SYSTEMS STUDIES

I have discussed only several of what are considered the most essential problems of systems research methodology and have attempted to summarize the results of their elaboration. It is certain that many other more special questions of methodology of systems studies, whose analysis occupies a prominent place in the literature have been left beyond the framework of this approach. Since it is not possible to deal with these topics, due to the limited length of this paper, I shall formulate in conclusion several of the most important problems of methodology of systems research which I feel now demand solutions. Naturally, a list of these problems can only be compiled on the basis of estimating the existing situation in this field, and therefore each item included is both a problem for future studies and to a certain extent the result of what has been achieved in this direction.

It would be appropriate to include in this list the following problems.

(1) *Making definitions more precise and constructing formalized descriptions of the basic concepts of systems research methodology.* What is meant in this case is such concepts as "system", "element", "structure", "wholeness", "connection", "hierarchy", "system-environment relation", etc. There has been a great deal of endeavour in this area, and quite naturally so, since any attempt at developing a systems approach or constructing a general systems theory sooner or later will necessarily come up against this problem. But despite this, however, at present there is no system of basic concepts of systems studies which is more or less generally accepted. One only has several approaches to the construction of such a system. Usually, they are quite isolated from each other, the same systems concepts are defined in

essentially different ways, the definitions are not rigorous. This is why this task is extremely urgent.

(2) *Theoretical description of specific methods of systems research* that distinguish one form of scientific and technological activity from other types and forms of scientific study (substrate, genetic, functional, structural, etc.). Despite the fact that the formulation of the problem itself is definitely not new, it seems that no serious attempts to solve it have yet been made. For this reason, as far as their methodology is concerned, systems studies appear as an agglomeration of a variety of most diversified methods, procedures and techniques (at one extreme, of purely empirical nature, associated with attempts to investigate specific material systems, and at the other – as abstract mathematical models in the development of the mathematical systems theory) in which neither their special systems content nor their position and functions in the system of systems research methods have been clarified. The latter are absolutely non-existent, this is why the development of the above problem is necessary.

(3) *Analysis of peculiar epistemological features of systems research.* The urgent need for dealing with this problem stems from the fact that as the object as a system is studied or designed, many classical epistemological topics, such as the object-subject relation in knowledge, the connection between the truth and correctness of the decision made by the subject, etc., acquire new meaning. This is primarily because of 1) the high degree of complexity of typical systems problems whose solution is usually only plausible; 2) the more important (compared with the classical scientific studies) role of the subjective factor in the elaboration of systems problems; and, 3) because of the fact that the process of systems investigation is internally contradictory; this is expressed, for example, in the presence of systems thinking paradoxes (see [33, pp. 232-246]). References [13, 28, 52, pp. 434-443] reveal several essential epistomological features of systems studies; in this field, however, there is still a great deal to be done.

(4) *Comparative analysis of conceptual structures and the fundamental principles of systems approach (the general systems theory) and the set-theoretical methodological concept.* This task is closely related to problems 1-3, and can be regarded in a certain sense as their concretization. The reason I single it out as a special problem is that many aspects of the opposition "systems – set-theoretical research" are decisive both for understanding the essence of the systems approach and for constructing the specific systems conceptual apparatus. Indeed, classical science of the nineteenth and first half of the twentieth century is based on set-theoretical abstractions and principles, while science and technology of the second half of the twentieth century, which have adopted the methods of systems research, are attempting to overcome the limited nature of set-theoretical methodology. This widely recognized intuitive opposition, however, must be expressed in a systematic and rigorous way. The first steps in this direction have been undertaken in [41, 44].

(5) *Construction of the general systems theory as a systems metatheory.* The significance of the development of this methodological trend has been discussed in sufficient detail previously.

(6) *Creation of the theory of systems classification.* The nature of this kind of theory can be only methodological, while its results are equally necessary both for the development of systems research methodology and for the specialized-scientific analysis of specific (material as well as ideal) systems.

(7) *Elaboration of methodological foundations of simplification theory.*
Since those systems whose study is interesting both from theoretical and
from practical angles are extremely complex (the living organism, the human
brain, social organizations, industrial conglomerates, control systems, etc.),
the only possible strategy for their study and/or design is operation with
simplified models of these systems. Although a great many systems
theoreticians emphasize the importance of formulating a theory of systems
simplification, no substantial results have yet been obtained.

(8) *Development of methodological foundations of the theory of hierarchical
systems.* This branch of general systems theory has recently been the object
of growing attention. M. Mesarovič and his co-authors [27] who deal with
several essential characteristics of any hierarchical organization made a
certain contribution to the construction of the theory of hierarchical,
multilevel systems. Further development in this field has been greatly
retarded by the lack of sufficient methodological foundations for such a
theory.

(9) *Study of methodological problems of systems analysis.* As was noted
above, systems analysis is now the most extensively used means of systems
research for solving practical problems. So obviously the efficiency of this
means depends primarily on how profoundly its methodological nature is
understood. Nappelbaum [28] suggests an interesting programme for
elaborating methodological and theoretical problems of systems analysis.

(10) *Analysis of the possibility of applying the methodological ideas and
principles of systems research in solving specific scientific and
technological problems.* Methods of approaching the requirements posed by the
modern scientific and technological revolution, such as global modelling, the
creation of automated control systems, the development of complex computer
networks, the solution of the ecological problem, etc., more than traditional
scientific and technological problems should be in the focus of researchers'
attention. The above problems are tackled first of all by the methods and
techniques of systems analysis.

(11) *Construction of a methodology of science based on the principles of
systems approach and the general systems theory.* No matter how paradoxical
it may be, the systems approach, representing one of the scientific and
methodological trends in modern science, has until now had practically no
influence on the construction of a methodology of science. And this (it must
be emphasized that the point in question is not philosophical methodology but
specialized scientific and methodological concepts) is now in a period of
serious restructuring expressed primarily in the transition from the
hypothetical-deductive substantiation of scientific knowledge ideas to
historical and methodological models of the structure and development of
science (see [19, 38]). Such models, e.g. Kuhn's paradigmatics, I. Lakatos's
methodology of research programmes [23, 38], and a number of models of
scientific evolution suggested by Soviet authors (for instance [14, 37]),
explicitly or implicitly employ the systems approach principles both for an
understanding of their objects (a scientific theory, a historical sequence of
scientific theories, a scientific community, etc.) and for the construction
of a theoretical description of their "activity" (the establishment of the
integrity of scientific knowledge, the analysis of the relation between
philosophy and specialized scientific knowledge, elucidation of systems
connections between the empirical and theoretical levels of knowledge, etc.).
The problem is to turn these systems intentions which are scattered and not
always clearly formulated in the interpretation of the methodology of science
into a rigorous scientific theoretical construction (see [31, 34]).

(12) *Application of systems approach principles to the creation of the science of science.* This requirement stems from the realization of the systems nature of the science of science, the latter discipline amounting to the complex study of all essential facets and aspects of scientific and technological activity, the organization of modern science and technology, and their control. Although there are already some results in the analysis of this field (see, for example, [19, 36]), as far as its general (methodological and scientific-theoretical) aspects are concerned, the above problem is a long way from being definitively solved.

These, as I see them, are the main tasks of systems research methodology, whose development is now of the greatest interest. In conclusion I would like to point out that I have dealt here only with the problems of systems study methodology, rather than with the topical questions of systems studies as a whole. These definitely do not coincide with the list presented above. It is equally obvious that some more general issues from this field recently in the focus of attention should be analysed further, and, first of all, the individualizing of the relationship between the philosophical methodology of dialectical materialism and the systems approach, as well as the investigation of the history of systems-methodological concept development in science and technology, plus the problems of systems research methodology formulated above.

REFERENCES

1. A. N. Averyanov, *The System: Philosophical Category and Reality*. Moscow (1976), in Russian.
2. R. L. Ackoff and F. E. Emery, *On Purposeful Systems*, Chicago (1974).
3. V. G. Afanasyev, *The Problem of Integrity in Philosophy and Biology*. Moscow, Mysl' (1964), in Russian.
4. V. G. Afanasyev, On the systems approach in social cognition, *Voprosy Filosofii* 6, 98-111 (1973), in Russian.
5. I. V. Blauberg, V. N. Sadovsky and E. G. Yudin, *The Systems Approach: Premises, Problems, Difficulties*. Moscow, Znanie (1964), in Russian.
6. I. V. Blauberg, V. N. Sadovsky and E. G. Yudin, The systems approach to modern science, in: *Methodological Problems of Systems Research*. Moscow, Mysl' (1970), in Russian.
7. I. V. Blauberg and E. G. Yudin, *The Development and Essence of The Systems Approach*. Moscow, Nauka (1973), in Russian.
8. I. V. Blauberg, The systems approach as a subject of historical-scientific reflection, in: *Systems Research Yearbook, 1973*, pp. 7-19 (1973), in Russian.
9. I. V. Blauberg, V. N. Sadovsky and B. G. Yudin, The systems principle in philosophy and systems approach, *Voprosy Filosofii* 8, pp. 39-52 (1978), in Russian.
10. A. A. Bogdanov, *General Organization Theory, Tectology*, Vols 1-3, 3rd ed. Moscow (1925-1929), in Russian.
11. W. Gasparski, *Kryterium i metoda wyboru rozwiązania technicznego w ujęciu prakseometrycznym (z zagadnień metodologii projektowania)*. Warszawa, Panstwowe wydawnictwo naukowe (1970).
12. D. M. Gvishiani, *Organization and Management*. Moscow (1972), in Russian.
13. D. M. Gvishiani, Materialist dialectics as a philosophical basis for systems research, this book.
14. B. S. Gryasnov, On the interrelations of problems and theory, *Priroda* 4, 60-64 (1977), in Russian.
15. S. I. Doroshenko, Scientometric indicators of Soviet literature on systems research, in this book, pp. 127-135. Moscow, Nauka (1978), in Russian.

16. S. V. Emelyanov and E. L. Nappelbaum, The fundamental principles of systems analysis, in: *Problems of Management in Socialist Industry*. Moscow, Economika (1974), in Russian.
17. *Studies in General Systems Theory: Collection of Translated Papers* (V. N. Sadovsky and E. G. Yudin (Eds)). Moscow, Progress Publishers (1969), in Russian,
18. D. Cleland and W. King. *Systems Analysis and Project Management*. McGraw-Hill (1968).
19. *Communication in Modern Science:* Collection of translated papers (E. M. Mirsky and V. N. Sadovsky (Eds)). Moscow, Progress Publishers (1976), in Russian.
20. V. I. Kremyansky, Some specific features of organisms as "systems" from the point of view of physics, cybernetics and biology, *Voprosy Filosofii* $\underline{8}$, 97-107 (1958), in Russian.
21. V. I. Kremyansky, *Structural Levels of Living Matter: Theoretical and Methodological Problems*. Moscow, Nauka (1969), in Russian.
22. V. P. Kuzmin, *The Systems Principle in Marx's Theory and Methodology*. Moscow, Politizdat (1976), in Russian.
23. T. S. Kuhn, *The Structure of Scientific Revolutions*. Chicago, University of Chicago Press (1970).
24. V. A. Lectorsky and V. N. Sadovsky, On the principles of systems research (in relation to L. Beralanffy's "general systems theory"), *Voprosy Filosofii* $\underline{8}$, 67-79 (1960), in Russian.
25. V. A. Lectorsky and V. S. Shvyrev, Philosophical methodological problems of the systems approach, *Voprosy Filosofii* $\underline{1}$, 146-153 (1971), in Russian.
26. A. A. Malinovsky, *Trends in Theoretical Biology*. Moscow (1969), in Russian.
27. M. Mesarović, D. Macko and Y. Takahara. *Theory of Hierarchical Multilevel Systems*. New York (1970).
28. E. L. Nappelbaum, Systems analysis as a research programme: guidelines and objectives, this book.
29. L. K. Naumenko, Hegel's dialectics and the systems approach, *Filosofskie Nauki* $\underline{4}$, 95-103 (1974), in Russian.
30. A. P. Ogurtsov, Stages in the interpretation of the systems principle of scientific knowledge (Antiquity and Modern time), in: *Systems Research Yearbook, 1974*, pp. 154-186. Moscow, Nauka (1974), in Russian.
31. A. I. Rakitov, *The Philosophical Problems of Science: The Systems Approach*. Moscow, Mysl' (1977), in Russian.
32. V. N. Sagatovsky, A study in the development of conceptual categories for the systems approach, *Filosofskie Nauki* $\underline{3}$, 67-78 (1976), in Russian.
33. V. N. Sadovsky, *Foundations of General Systems Theory: Logico-Methodological Analysis*. Moscow, Nauka (1974), in Russian.
34. V. N. Sadovsky, Methodology of science and systems approach, in: *Systems Research Yearbook, 1977*, pp. 94-111. Moscow, Nauka (1977), in Russian.
35. M. I. Setrov, *General Principles of System Organization and Their Methodological Significance*. Leningrad, Nauka (1971), in Russian.
36. *Systems Research Yearbook*. Moscow, Nauka (1969-1982), in Russian.
37. V. S. Stepin, *The Development of a Scientific Theory*. Minsk (1976), in Russian.
38,39. *The Structure and Development of Science, Boston Studies in The Philosophy of Science:* Collection of translated papers (B. S. Gryasnov and V. N. Sadovsky (Eds)). Moscow, Progress Publishers (1978), in Russian.
40. A. I. Uemov, Methods of construction and development of general systems theory, in: *Systems Research Yearbook, 1973*, pp. 147-157. Moscow, Nauka (1973), in Russian.
41. A. I. Uemov, *The Systems Approach and General Systems Theory*. Moscow, Mysl' (1978), in Russian.

42. Yu. A. Urmantsev, *The Symmetry of Nature and The Nature of Symmetry*. Moscow, Mysl' (1974), in Russian.
43. K. M. Khailov, The problem of systems organization in theoretical biology, *Zhurnal Obshchei Biologii* 24, 324-332 (1963), in Russian.
44. Yu. A. Shreider, Set theory and systems theory, in: *Systems Research Yearbook, 1978*, pp. 70-85. Moscow, Nauka (1978), in Russian.
45. G. P. Shchedrovitsky, *Methodological Problems of Systems Studies*. Moscow, Znanie (1964), in Russian.
46. G. P. Shchedrovitsky, *The Systems Approach and Trends in the Development of Systems-Structural Methodology*. Obninsk (1974), in Russian.
47. E. G. Yudin, *The Systems Approach and The Principle of Activity: Methodological Problems of Modern Science*. Moscow, Nauka (1978), in Russian.
48. *Basic and Applied General Systems Research: A Bibliography* (G. J. Klir and G. Rogers Eds in cooperation with R. G. Gesyps). State University of New York at Binghamton, New York (1977).
49. D. Berlinski, *On Systems Analysis: An Essay Concerning the Limitations of Some Mathematical Methods on The Social, Political, and Biological Sciences*. Cambridge, MIT Press (1976).
50. L. von Bertalanffy, *General System Theory: Foundations, Development, Applications*. London, Penguin (1971).
51. P. Caws, Science and system: on the unity and diversity of scientific theory, *General Systems* 13, 3-12 (1968).
52. G. J. Klir (Ed), *Trends in General Systems Theory*. New York, Wiley-Interscience (1972).
53. E. Laszlo, *Introduction to Systems Philosophy: Toward a New Paradigm of Contemporary Thought*. New York, Gordon and Breach (1972).
54. R. Mattessich, *Instrumental Reasoning and Systems Methodology*. Dordrecht, Reidel (1978).
55. P. K. McPherson, A perspective on systems science and systems philosophy, *Futures*, June, pp. 219-239 (1974).

SYSTEMS ANALYSIS AS A RESEARCH PROGRAMME: GUIDELINES AND OBJECTIVES

E. L. Nappelbaum

The current situation in scientific research is reminiscent of what occurred in science at the turn of the century. Today we all know that many of the research ventures which emerged at that time led to major breakthroughs and dramatically altered the very outlook of science. But at that time, their pioneering role was not in the least obvious, and their most prominent feature was their intrinsic strangeness, the fact that they were in discord with the mainstream of scientific development, and that it was even difficult to see that they were valid in the light of scientific standards and values prevailing then.

Of course, it would be rather premature to claim that the newly emerging fields of studies will have a similar impact in the future, but the strangeness certainly does exist. Without any special effort one can now identify several recent research directions which have diverted from the traditional path of scientific progress and which still lack credibility as truly scientific research. But the current situation contributes at least one other new dimension. The pioneering research of the outset of the century did not alter the principal structure of scientific knowledge, neither did it cut across interdisciplinary boundaries; rather, it expanded them. The ideas of quantum mechanics or the theory of relativity were certainly revolutionary, but at the same time they simply created a new level of understanding, a more profound layer in the overall pattern of scientific knowledge which could be easily fitted in to the already existing structure.

In contrast, most of the out-of-the-ordinary modern research endeavours claim to be something highly original conceptually, and place themselves at a certain distance from everything else. It is even difficult to present their substance in terms of traditional concepts and notions.

Systems analysis appears to be one of the research ventures in this latter category.

Discussing the content and history of systems ideas is a formidable and hazardous undertaking. Systems analysis is in a constant state of brewing and fermentation, which is perhaps why the number of known explications of the meaning of systems analysis as a concept is in the same order as the number of people working actively in the field, to say nothing of the vagueness of the ideas of those who are simply tracing the developments in systems analysis, anticipating the moment when it becomes truly operational and is able to be readily applied elsewhere. Possibly the only thing that can be said about systems analysis without risking involvement in lengthy and rather meaningless discussions is that it has not yet reached the stage of genuine maturity and cannot be considered a field of knowledge or a package of scientific skills with established boundaries and an inner structure. At the same time, one senses indications of advancing stagnation, and there are voices either from the outside or from the sidelines of systems analysis which have been quick to claim that it as a whole can go no further, and that in the final analysis it has proved itself conceptually unproductive and devoid of originality.

Obviously the methodology of such a field of knowledge cannot be developed on the basis of post-rationalization. This might provide the explanation of why so much effort is spent to link systems analysis to some prior solidly established ideas, and to make the point that it in fact contains nothing revolutionary and that its prime merit is simply that it makes certain ideas explicit. On the other hand, there is another approach which says that it is in essence something conceptually new, and which attempts to construct the understanding of the meaning of systems analysis around these seeds of innovation. No matter what, every attempt to explicate what systems analysis is and can do should always be considered as a forecast, be it explicit or implicit, of how it will evolve in the future. In fact, systems analysis is a kind of research programme which outlines certain objectives to be pursued and establishes certain guidelines that allow us to assess what current trends of systems analysis development are important and what others might be misleading. This is the purpose of this chapter.

Of course one can only hope that this kind of programme turns out to be of some objective value and does not simply reflect the personal idiosyncrasies of its author. However, the principal objective of every forecast is not only to provide an accurate picture of the future but to a greater degree to highlight some hitherto unnoticed issues, to present new challenges, and to encourage imaginative thinking. That is why I hope that the programme outlined in this chapter will be construed not as an attempt to foresee the future of systems analysis or (even worse) to manipulate it, but as a call for further quest.

As a matter of fact, new efforts to evaluate the field as a whole appear to be rather current. There have been too many instances of late when original research based on solid substantial grounds and seemingly opening up new horizons has been buried under an avalanche of companion papers of a totally diversionary nature and having nothing at all to do with the substance of the subject matter at hand. To a certain degree, this evaluation can be applied equally well to some of the dimensions of the development of systems analysis. So if one wishes to avoid the supposedly imminent stagnation alluded to above, one must clarify one's general ideas and inherent system of values.

To forecast the future at the turning point of evolution is definitely not a simple task. At such moments one never knows which currents will persist and which will be changed, and thus what may serve as an authentic foundation for the projection. So in trying to construe its future significance we are not planning to base ourselves on the past experience of systems analysis and are not thinking of delving into the intricate pattern of the interrelationship between systems analysis and other sciences and philosophy. This, however, is not to say that we question the validity of this kind of macroscopic

approach to the examination of the content of systems analysis, systems approach and systems thinking, as represented for instance in [4, 8]. Nonetheless our present objective is more humble: on the basis of numerous and sometimes rather irrelevant (if considered in their own right) facts of current scientific life, we will try to identify some actual requirements in certain changes and anticipate their consequences. This approach will, one hopes, bring us closer to the practical considerations of those now working in the field and who will eventually be responsible for what will materialize as systems analysis in the future. Or to put it in another way, we will be examining systems analysis in the narrow sense of the term.*
This will also save us from brooding over terminological niceties and allow us to employ terms "systems analysis", "systems approach" and the like interchangeably, which to a certain extent goes against the grain of the emerging tendency to categorize these and other related concepts in a hierarchical manner.

SYSTEMS ANALYSIS AS A NOVEL DIRECTION OF SCIENTIFIC RESEARCH OF A NON-TRADITIONAL TYPE

As mentioned previously, the attempt to expound the concept of systems analysis will be based on the experiences of a variety of non-traditional research ventures which have been appearing in scores over the past few decades. Management science, control theory, cybernetics, operations research, problem solving theory, action theory, artificial intelligence, knowledge representation theory, data theory, some of the aspects of mathematical psychology, praxiology, systems engineering, some of the branches of normative economics and sociology, decision-making theory, etc., can be quite instructive in this respect. I am certain that each reader interested in systems analysis can extend this list *ad infinitum* along lines consistent with his or her acquaintance with the field and personal inclinations. An interesting analysis of some of these novel disciplines may be found in [5]. Some of the new fields of studies fell into oblivion even before they were voiced, and are remembered by only a minute number of specialists who refer to them in footnotes. Others have been inflated by an impressive number of publications and are now iron-clad examples of academic respectability. Other differences are also apparent. Nevertheless, there are some very important features in common, as have become evident even from a superficial examination.

First, each of these research issues has a non-traditional and what could even be called elusive object of studies. One may even go so far as to say that the main objective of these studies is not to acquire knowledge about an object or phenomenon, but rather to develop methods of solution for a certain class of problems. This emergence of methods as the primary object of scientific research sets all these studies completely apart from the traditional sciences. (A similar viewpoint is expressed in [3] and [10].) Indeed, even mathematics deals with the properties of quite real (though abstract) objects, and only indirectly with the methods making the studies of these objects possible.

---

*A more comprehensive examination (along lines consistent with the general ideas presented here) is offered in [4] and [12].

Another point which becomes noticeable when trying to trace the history of the evolution of each of these ventures (provided their evolution was not halted right at the very outset) is a definite sense of ultimate frustration. The impression is that those who might be called the midwives of scientific research were undertaking something very different from what we have now. Even in those instances when the evolution of the field has been dictated by meaningful considerations, there is still every reason to believe that it is not the meaning which basically motivated the pioneering research and not the meaning which guided our own expectations when we decided to study the field.

Finally another common feature of all these research ventures is in the tendency for them to break apart into several narrower and more specific subjects, and for extensive and challenging research objective to be replaced by limited and rather modest objectives with no claims to anything spectacular (rather this line of inquiry may also be feasible in some case). As a result, those working in the field have even begun to stop using its original name - which now appears somewhat pretentious - and explain their activity by reference to the limited specific problem they are dealing with.

One of the main assumptions of this chapter is that all this research should not be regarded as a collection of isolated and independent endeavours, but rather as different attempts to find a correct answer to a definite (though not yet fully appreciated) social requirement for the development of a novel research discipline. Thus it should come as no surprise that many pioneers in this non-traditional area of scientific research migrate from one field to another as if repeatedly trying to grasp something which escapes them to find the most appropriate "name-tag" for this "something". The day before yesterday the most appropriate name-tag for their efforts was cybernetics or operations research, yesterday it happened to be management science, today it is systems analysis, tomorrow - probably artificial intelligence or something else. Recent occurrences have thrust systems analysis to the forefront of these quests, and it will remain there only to the extent to which it meets public expectations underlying this area of research.

But why is a new scientific approach or a new scientific discipline required? And how does one go about revealing the essence of these new requirements? In our opinion, a search of this nature should be aimed at *identifying a new class of objects*, objects which are impossible or simply too difficult to study by traditional techniques, and which have also come to play such an important role in contemporary life that there is an urgent need to find a way of analyzing them. This intrinsic orientation of the search towards a specific class of objects may be quite valuable from several points of view. First and most important, it provides a solid foundation for our methodological speculations by bringing them closer to the very real needs of those having to deal practically with these objects. At the same time it makes it easier for those engaged in this area of endeavour to comprehend our interest in various theoretical issues and verify the consistency of the theoretical formulations and assumptions by the requirements of practice. Finally, this kind of approach presupposes right from the outset the limits of possible application, which is one of the necessary prerequisites of any mature science and is all too often disregarded by leading proponents of systems concepts.

Moreover (and probably this is the most important factor of all) the search for this new class of objects does not seem an overwhelmingly difficult one. As a matter of fact, all experience of the decades quite clearly indicates the emergence of such a class.

The most appropriate term to identify this novel category of objects, if one confines oneself to a single word, is "complex". Complexity has recently become one of the major topics in various research efforts (both of the traditional and non-traditional type). However, complexity is an extremely vague and ill-defined concept, and if one would like to employ it as the basis for defining the class of objects, one has to indicate explicitly in what rigorous sense it is going to be used. In this chapter I will call an object complex if it meets at least one of the following requirements.

(1) For an object of this class, it is *impossible* or *unfeasible* (for in this instance, the practical possibility of analysing it or of interpreting the subsequent results vanishes) to conceive a *unique model* (unique theory, unique explanation) which will be equally suitable for analysing every aspect of its behaviour from some uniform vantage point.

(2) Each of these individual uniform aspects of its behaviour is not inherently monodisciplinary with regard to the established disciplinary pattern of scientific knowledge but rather demands that allowances be made for different factors traditionally studied by *different* scientific disciplines.

(3) To interpret its behaviour, one has to take the *purposefulness* of the behaviour of some of its components into consideration.

(4) For an object of this class, there exist one or several decompositions in regard to which the purposes or functions of the different components are stable and at least in principle may be identified, but nevertheless the goal or function of the object *as a whole* is *unknown* or *underdefined* and in the final analysis depends on the way in which the different components will choose to pursue their objectives and on the interaction pattern that will subsequently emerge.

This definition obviously needs further elaboration.

First of all, it has to be underlined that a multiplicity of models to represent an integral "physical" object is also quite common to the traditional sciences. Even such a "simple" object as a stone can be studied by science as a massive point, as an absolutely rigid body, as a mineral, as a chemical compound, as an insulator, etc. In the above definition the crucial role is played by the concept of "uniformity". Thus traditional science has always considered each model as suitable for different isolated moments, conditions, "plans" or "forms" of existence which are very precise in regard to this decomposition. The idea of the controlled experiment is itself based on the hypothesis that we can provide those conditions for experimentation which will guarantee that only one of the models applies, and that the number of factors involved is relatively low. Moreover, it assumes that events of the "everyday" life of the object may be interpreted through references to these monomodel theories without any special extra effort.

But when the phenomena become more complex (the number of variables increases) or involve the border zones of the relevance of different models, the situation changes dramatically. In this event, no single "pure" model can credibly account for the behaviour of the object by itself. Meanwhile, mechanistically piecing together the different models into a single supermodel does not answer the question either. First, because the computational difficulties of analysing its behaviour make it extremely hard to establish all the consequences of the changes in the endogenous variables (this side of the complexity curse was amply commented upon by R. Ashby [1]). But even more important is the cognitive complexity of this kind of super-model, which prevents from meaningfully interpreting the ascertained changes in each of the numerous variables of the model resulting from the significant variations of endogenous variables even if we were able to do so. This makes it impracticable to represent system behaviour by a single model. Instead, the only solution is to bring a range of mutually complementary models into the picture, each of which is inherently incomplete and is not able to present the excuse of overlooking something inessential to overall behaviour. Thus each individual model is not valid for object behaviour analysis, but validity may be ascertained only by a complete set of models.

Deprived of the traditional monomodel approaches to the scientific research of many important phenomena, the different non-traditional research ventures alluded to previously have developed both implicitly and explicitly their own approaches which operate not on the basis of specific features of the object being studied, but on the basis of specific interactions that one would hope to see it involved in (goal-oriented relativity of object representation), and specifically, on the basis of specific schemata (paradigms) of scientific explication which will open a way for meaningful interpretation in terms of feasible interactions [9]. This approach requires cutting across interdisciplinary boundaries. Multimodality of representation made it necessary to look for models which in the framework of a single border zone allows for very different factors, factors which are relevant to the same interaction. In other words, the answer to the problem of explaining the behaviour of the novel-class objects because of the impractibility of a single model development lies not in the projection of this behaviour onto the "planes" of different scientific disciplines, but rather in the design of non-trivial viewpoints to enabling the elucidation at the same time of both the main features of the planned interaction and the principal attributes of the object involved, which may be associated with different traditional disciplines.

The great majority of the objects of this type consist of several interacting goal-seeking components. This is obviously true primarily of objects consisting, among other things of human beings, whose individual and collective behaviour essentially governs the behaviour of that object. Whenever we are obliged to consider objects with human beings or other goal-seeking elements as components, we are compelled to alter our entire scientific outlook. We are forced to contemplate the functioning of the object not solely through the prism of the research worker's objectivized cognition, but also by making allowances for the perception of reality by those who govern the behaviour of the object and who react to the environment in a manner consistent with their objectives, opinions, beliefs, preferences and values. To allow for this possibility of equivocal apprehension, not only are certain specific new techniques required, but one also has to change one's entire thinking about established scientific concepts. Specifically, one has to discard the traditional concept of the controlled experiment since the problem of translating inferences gained through enforced unambiguous comprehension of the experimental conditions in the framework of field behaviour happens to be in no way simpler than the original one (this becomes quite apparent from the experience of experimental psychology). In other words scientists dealing with novel kinds of objects must not only evolve adequate viewpoints to cover all significant aspects of the object's behaviour, but try to anticipate the viewpoint of the object's components as well.

Of course, the first two conditions alluded to above certainly are not only
met by the "human" objects of study. They are equally valid for many
"artificial" (technological) objects which bear the indelible impressions of
the goals pursued by the designers. So in order to be able to deal
effectively with the explanation of the behaviour of these objects, one has
to isolate aspects portraying objective relationships and those which can be
attributed to the desires of the human beings who have designed it. A
similar situation arises in analysing the behaviour of various complex
biological objects. In this case, the teleological description of the
functioning components is often the only possibility of reducing the
complexity of the integrated comprehension of the entire object behaviour.

The goal-seeking nature of some of the system components leads to basic
changes in the meaning of many general methodological concepts. In
particular, the content of the concept of structure has to be re-evalued,
since for the objects of this type all structural (and not only
structural) interactions between different components are shaped not "from
top to bottom" (i.e. not predetermined by the holistic aspects of the
object), but rather "from bottom to top" (i.e. in accordance with the
choice made by the components themselves as they pursue their goals).
This is specifically the core of the difference between the so-called ill-
structured problems of non-traditional research, and the well-structured
problems of traditional science.

As mentioned previously, all these characteristic features of the novel
objects are not mutually independent. Meanwhile in different objects of this
type, only some of them are immediately explicit, while the other less
prominent features can be detected only with some effort.

To simplify the following, let us agree to call *systems* those objects meeting
the above requirements, with the proviso that we are using the term in its
narrow sense and that as such, it is not applicable to a wider category of
objects which were called systems in ordinary language, and in some systems -
scientific and engineering disciplines.

Today many systems and objects of the type characterized above are known,
have attracted a great deal of attention in science and practical
applications and are the subject of a vast range of studies.

But what has made them so essential, and why previously was it not so
imperative that they be studied? Of course, many such systems, e.g. global
systems, regional and scientific development systems, urbanization systems
and other socio-economic systems and organizations can be said to have always
existed in one form or another. However, influenced by numerous socio-
economic factors, particularly by scientific and technological revolution,
quite recently the intensity of various interrelationships within these
systems has greatly increased. And this has affected their dynamics in such
a way that many control and management problems which were able to be solved
locally through the adjustment of partial aspects of their behaviour now
require an integrated holistic approach.

Another important reason why systems studies are so urgently needed is that
the vast majority of the objects of this type are of artificial nature, i.e.
created through the (conscious or unconscious) purposeful action of human
beings. One of the principal human "design" strategies is the drastic
modification (usually amplification) of natural interaction intensities, and
because of this, artificial systems have very rapidly extended to the
"dangerous" boundary regions of their internal interactions. The result is

that along with the properties which human "designers" purposefully sought, these systems have acquired other features linked to the unpredictability of boundary interactions, and thus not anticipated even by the creators of the system themselves. This observation applies not only to socio-economic systems. For instance, the same sort of effect may also become essential in sufficiently complex technological systems, as illustrated by such rigidly pre-programmed systems as software systems.

Another interesting phenomenon resulting in the abundance of systems as an object of studies is linked to the internal dynamics of science. Indeed, in various scientific disciplines, the process of acquiring increasingly profound scientific knowledge has now come into contradiction with the limited interpretational cognitive capacity of human beings as potential users of scientific knowledge. The persistent search for further detailization and deepening of knowledge, combined with the accumulation of highly specialized specific scientific information, leads to a point at which it becomes impossible for a human being to reconstruct some meaningful pattern out of them. Gaps emerge which prevent our understanding of systems behaviour at the micro- and macro-level (holistically). This breakdown becomes manifest not only in science (in biology, for instance, where this situation is particularly evident) but in the humanities as well, where there is also an overproduction of narrow specialists and a shortage of generalists, i.e. people who understand general global holistic patterns of integrated behaviour.

Many scientific disciplines and specialized theories claim to surmount this hurdle. However, none of them appears to have been able to pinpoint the core of the problem (even for the limited framework of a specialized viewpoint). This is probably the reason why many of the results seem to be of no importance whatsoever, and have neither practical implementation nor natural follow-up.

The experience of these attempts may serve as a catalyst for subsequent deliberations. Today there are sufficient reasons to assume that the ultimate collapse of the research attempted was preordained by several factors. First of all, none of these theories has so far embarked on a study of the objects of the new type (systems) in an integrated fashion. They have chosen to focus their research endeavour on some very specific albeit very important aspect of systems behaviour. Second, these theories have attempted to remove human beings (regardless of the capacity in which they are involved - system component, observer or designer) from the realm of the research. As a result, several genuinely "human" notions (for instance, the concept of "purposefulness") were considered in an over-absolutized manner.

Systems analysis is still free from these deficiencies. Its results have not yet been absolutized, and because of the real requirements of everyday practice, its objects are studied in an integrated global manner, although not always consciously.

However, this alone does not provide enough grounds to claim that systems analysis will continue to play the leading role in the quest for new knowledge and novel types of understanding. To be ultimately successful, systems analysis must clarify the main objectives of its own development, establish consistent guidelines and teach itself to avoid the numerous pitfalls awaiting it as it develops. Systems analysis, which is used today as a rather primitive tool in the goal-oriented (from the viewpoint of feasible interactions, and mainly of control) study of complex objects, should ultimately evolve into a body of sophisticated knowledge and skills related to the understanding of the behaviour of objects which we choose to call systems.

CRUCIAL SYSTEMS ANALYSIS DEVELOPMENT PROBLEMS

It is more or less contingent that system analysis in its current shape has been pushed into the forefront of the scientific onslaught which will provide us with the tools for the study of complex systems. That is why it is only natural to ask what makes us count so highly on the development of systems analysis rather than try and reconceptualize one or several of the theories which, as we claim, have been unable to solve a similar problem or elaborate another original direction of scientific research. First of all, this is because the leading role of systems analysis is qualified only from the chronological perspective. We are convinced that irrespective of the degree to which cybernetics, operations research, management science or any other scientific theories hitherto mentioned has been successful, scientific development sooner or later was bound to prompt to the statement of the problems which are today the object of study (sometimes in an incredibly naive manner) by systems analysis.

The point is (and this is another salient feature of the present stage) that even before systems analysis becomes an accepted scientific discipline, it has to endure and evolve under conditions in which society anticipates the need to apply its results and techniques even if insufficiently developed and validated, and cannot postpone the solution of the appropriate problems for more opportune times. This is the source of both the strength and weakness of systems analysis. The strength - because it makes systems analysis constantly aware of the invisible guiding hand of practice, unceasingly extends the range of "proper" research objects, and makes it impossible to disregard real human needs or to "factor them out" without risking serious damage. The weakness - because frequently it leads to immature solutions, to sweeping existing difficulties under the carpet and to a willingness by certain people coming to the field to use its colours without accepting any "systems" obligations. Taking all this into account, we should try to sketch a systems analysis development programme based on practical requirements and the accumulated experience of systems studies. In accordance with this plan, we will initially need to search for novel problems to tackle in the course of systems studies as we see them now or may conceptualize them ideally in the future.

Diagnostic or *context engineering* problems constitute the *I type* of problems of this type. The need to design a certain level, plan or system of object interactions with the environment which should provide us with the scope of the models to be elaborated prompts the necessity of establishing some (fuzzy in general) boundaries which will separate the system from its environment and to some extent will define the limiting depth of the interaction influence at which we are going to halt our analysis; of assessing the ways and means of the actual interaction; of integrating the system under investigation in a wider system both in the sense of it being a part (a subsystem) of a more complex "metasystem", as well as a specific instance of a class of more or less similar systems providing us with the problem "context" or problem area.

It should be noted that one of the primary differences between systems analysis and traditional scientific research is the analyst's far more active attitude to information obtained. Direct observations and superficial results should be regarded only as symptoms of something happening under the surface much in the way that we consider the advent of numerous non-

traditional research issues only as a symptom. Therefore it is not these
facts and observations (even if our client requests that we do exactly that)
which we must deal with, but rather the processes underlying them, and based
on the diagnosis of these processes, offer a reconceptualisation,
reinterpretation and revalidation of both the primary sources of information
and existing sources of new data. Problem area diagnostics is an integral
part of the specific viewpoint generation process, the process crucial to
systems analysis.

Systems analysis problems of the *II type* involve the development of
significant (promising in the instance of control action) interaction
strategies for the object being investigated, or (to express it differently)
the design of interaction options. Basically, we are now referring to the
necessity of searching in a general contextual framework for those specific
viewpoints that will define not just feasible strategies of interaction
with the object, but will offer research guidelines as well. To do this,
one has to investigate not just the object itself, but rather our own
experience of similar or related problem solutions, known interaction
patterns and properties of objects interacting with the object being studied
although strictly in the context of the interactions that thus emerge.

The detailed study of the object itself is carried out by dealing with
problems of the *III type*, when one needs to *develop a family of simulation
models* describing how each interaction will influence the object behaviour.
It has been already mentioned that systems analysis does not see this as the
creation of some integral supermodel encompassing every factor identified,
or every necessary aspect of each interaction discovered. Each of the
partial model offers answers, but only to a limited number of questions,
provides a particular rendering of observed phenomena and a different
structuring of the object, and as well as varying ideas as to which structural
peculiarities of the object should be retained (simulated) and which may be
discarded without losing the external "black-box" matching.

However, even if a sufficient number of these models are developed and
analysed, the problem of how to merge the various findings about system
behaviour into an integrated pattern still remains. This will be solved not
by the development of some kind of supermodel, but rather through the
analysis of the reaction of other interacting objects to the observed
behaviour (in control and management problems, the reaction of the decision-
makers and strategy planners). In principle, this leads to the study of
simply another "no-man's land" situated somewhere between the environment
and the object, but closer to the environment. It provides a basis for
meaningful insight into interaction contingency and pattern, and determines
its place in the larger system in which it serves as a component. The
problems of this *IV type* may be called *decision* and *planning model design*
problems, as they have to be dealt with at the management stage which is
usually referred to by this name.

Every systems study may be reduced to the development of various viewpoints.
An overall outlook can be formed at the problem area design stage, while
more specific standpoints are developed at the option and simulation model
design stages, only to be again aggregated, though from a different
perspective, at the stage of decision model elaboration. At the same time
one of the specific features of goal-seeking systems is its considerable
freedom of choice of external (and internal) interaction perception.
Therefore both the analyst and the interacting systems should allow for and
predict how the subsystems of the system being investigated will conceive
the reality. In the framework of a more specific management system analysis

problem, this leads to emergence of the problems of another *V type* - that is, problems of *organizational design* which should implement the desired reaction to the chosen interaction. At this stage, information which could be said to have been extracted from the system and combined with the information about our skills is injected back into the system to provide the desired modification and changes. This category encompasses not only the ordinary problems of control and management design, but also those of how to organize fruitful relations with the potential supplier of the basic data and how to present the results of the study in a way that will lead to their acceptance by the ultimate users, the final report being one of the instruments to modify the system in the desired direction.

Even this brief characterization of systems analysis problems reveals their close interrelationship, which makes it impossible to disentangle them one from another and to deal with them separately both in time and in regard to the staff allotted to work on each of the tasks. Moreover, to be able to cope with all these problems, the analyst or a group of analysts should be rather broadminded, and possess a vast scope of knowledge and an extensive range of tools and skills for scientific research. Much of this knowledge and many of the skills and tools may be borrowed from other scientific disciplines. To prevent systems analysis from engulfing all these specific (and sometimes very unspecific) scientific disciplines whose results and techniques it uses, it seems reasonable to classify under this heading only those results and methods that are equally revealing for every type of problem alluded to previously. Thus the above classification should serve as a kind of sieve for sifting the crucial systems analysis development problems from the adjacent and monodisciplinary problems. These considerations enable us to identify the following list of research problems whose solution would have an appreciable impact on the development of systems analysis methodology as a whole.

(1) The first group of systems analysis techniques which demand further development and special concern involves the problems of *formalization, acquisition, storing* and *handling* of empirical *data*. The point is that our conception of the content and essence of systems analysis suggests the utilization of information of a special type or, to be more precise, unconventional interpretation. One of the main requirements of scientific discipline in the traditional sense of the word is in the maximum objectivity of the original data, in the removal of the slightest traces of value judgement, and in the elimination of the effects of measuring devices. This is why traditional science has always tried to work at a relatively low level of description and to avoid aggregation.

In contrast, systems analysis should make broad use of generalized, synthetic information containing a judgemental component and transduced by the perception of those who interact with the system. In other words, while traditional science tries to deal with the objective characterization of the state of nature, systems analysis uses this information after it has been deliberately transformed into a specific representation of this state or, expressed differently, into knowledge about this state if it assumed that in any knowledge we have already sacrificed some of its reality, while offsetting this loss by the logical organization of what remains. The principal directions of development for this group of techniques are related to the progress of:

(a) a modern measurement theory which attempts to enlarge the range of abstract objects suitable for an empirical relation description and for identification of critical properties of empirical phenomena which determine the relevance of these representations;

(b) a knowledge representation theory which should reveal the inherent structure of concepts perceived as an integral whole, study the transformations of this structure that accompany the transfer from one "viewpoint" to another, and ascertain their relation both to the reality they reflect and to the orientation of the subject (the first steps in this direction have already been taken under the general heading of knowledge representation theory as part of artificial intelligence, and have led specifically to the recent development of the frame formalism [7]);

(c) investigation of the specific features of socio-economic information and development of appropriate review and collection techniques (e.g. field studies of sociology, content-analysis, etc.);

(d) the theory and techniques of judgemental data handling, i.e. the data which is provided solely by human beings and which has been tempered by subjective considerations and intentionally or unintentionally distorted through the influence of considerations of its perceived usage by a recipient;

(e) the theory and hardware of large-scale data banks with the developed associative structure making possible the storing of not only a great volume of data about the system under investigation, but also all "external" knowledge needed for an efficient search for new interpretations, novel viewpoints and concepts;

(f) a general theory of data aggregation based essentially not on the ideas of (parametric or non-parametric) statistics, but rather on data organized in a complex dynamic structure defining semantic relations between different concepts and data from various categories.

(2) Another important direction in which systems analysis is developing is related to the attempts to open new ways for the design of original decision options, creative strategies, unconventional representations and latent structures. In other words, we should be able to find the ways and means to *amplify* the *inductive capacities* of human intelligence, while essentially all efforts are now being channelled into the development of formal means of amplifying deductive reasoning capacities. Serious research in this area has begun but quite recently, and does not have a unified conceptual foundation. Nonetheless, some promising lines of pursuit may be mentioned here as well. It is worth pointing out the development of formal inductive logic, of morphological analysis and other structural approaches to new options design, of syntectics and group interaction for creative problem solving, as well as investigations of major paradigms of inquiry reasoning.

(3) The necessity of using this new type of the information and the advent of new means to represent it formally prompts the reconsideration of *modelling* methodology as well. Simulation models that preserve and highlight certain structural features of the object modelled play the most important role in systems analysis. Therefore systems analysis may look primarily to the development of:

(a) semiotic modelling theory allowing for judgemental variables, particularly for different interpretations of the same variable by different systems units;

(b) object-oriented hierarchical simulation systems containing various partial pseudo-independent submodels interconnected in an integrated whole through aggregation (disaggregation) and interpretation operators coupling various levels of the overall hierarchical structure;

(c) the statistical (Monte Carlo) branch of simulation modelling based on the assumption of the homogeneity of numerous problem factors, and which reached a certain maturity in the past;

(d) the very important, though underdeveloped, theory of model sensitivity to the qualitative assumptions forming the basis of modelled system presentation.

(4) The most developed and at the same time the most specific field of scientific research is that of dealing with the problematics of *decision-making, objective structuring, programming* and *planning*. One cannot complain that there is an insufficient number of papers in this field or a shortage of qualified people involved. Nevertheless, an unbiased review will also reveal here that too many results are really unconfirmed invention and lack a common understanding of both the essence of the problems involved and the means of coping with them. The major topics of this line of inquiry include:

(a) the development of decisions plans and programmes efficiency evaluation theory;

(b) solution of multicriterial decision-making and planning problem;

(c) the study of uncertainty, particularly for the case in which this uncertainty is not of a stochastic nature but stems rather from the vagueness of judgemental data and the intentional simplifications of systems behaviour presentation;

(d) the development of individual-preference social aggregation for the decisions concerning several sides playing an active role in overall systems behaviour formation;

(e) investigation of specific features of socio-economic efficiency criteria;

(f) the development of objectives structure and plan consistency checking techniques, as well as the search for the proper trade-off between action programme uncertainty and its readiness for readjustment prompted both by the acquisition of new environmental information and by the deviation of the completed programme from the expected results.

But even more important, a major reconceptualization of the goal structure, plan and programme functions and the clarification of their interconnections will have to be made.

(5) And last but not least, special attention should be given to *goal-oriented programming* techniques. In this respect the first priorities should be:

(a) the development of a multilevel management system theory;

(b) the analysis and design of newly emerging organizational forms and interaction patterns;

(c) the further development of the goal-oriented programming concept;

(d) the analysis of organizational change dynamics particularly the sociological and psychological aspects of organizational behaviour.

All the above lines of systems analysis methodology development clearly entail the handling of great volumes of information with a complicated inner structure unfeasible (and if feasible, then with a very low efficiency) without computers. Moreover, in systems research computers should not play merely a passive role. They should operate interactively with human beings, sharing many of the creative research functions. One could claim the same of the role allocated to the systems research analyst in his collaboration with the computer. It should not be restricted to the role of a user, programmer and operator. Human beings will also be engaged in the capacity of :

— information transducers (expert evaluation man-machine systems);
— couplers of submodels and units, as a source of interpretation and value propositions (semiotic models);
— search heuristics generators (inductive capacity amplification technique);
— sources of preferences and trade-off parameters (man-machine decision and planning systems);
— active components of organization models (business gaming).

To be able to do all this, a different approach to man-machine simbiosis has to be found and effective principles of man-machine systems design developed. An auxiliary role in these problems (but quite independent in other respects and quite important in its own right) is played by :

— the theory of complex situation description, which should enable the development of efficient interaction between a human operator and a complex simulation computer system, and in particular, facilitate human cognition and interpretation of simulation results;
— development of dialogue facilities to provide the means of man-machine conversation using a language close to the natural one.

The programme of systems analysis development as sketched above obviously does not claim to be complete or unique. However, I feel it encompasses the gist of the problem.

In order to be truly involved in the systems analysis methodology development as represented by the programme outlined above, it is naturally not enough merely to announce that you are working in one of the mentioned directions. The experience of sometimes formally very advanced and sometimes quite naive research efforts along any of these outlined development directions illustrates this point quite forcefully. At the current stage in the development of systems analysis when we are forced to proceed from the cursory requirements of the moment and are unable to wait for the creation of a solid scientific foundation, there is a great need of some dominating idea, of some general methodological basis, of a touchstone to perfect the conceptual framework for various techniques and skills, and for checking the consistency of the chosen direction with the overall spirit of systems analysis.

As we see it, the role of this kind of yardstick can be played by various considerations related to the explanation of human individual and collective purposefulness. Of course we would be quite pleased to replace this vague allusion to some "considerations" by a definite reference to a theory of human purposefulness developed on the basis of empirical observations of various forms of human rational goal-oriented behaviour. However, such a theory has not yet been developed, and we are aware only of some isolated attempts (e.g. [2, 11]) to create a foundation. Meanwhile, everybody does have a certain understanding of his own and others purposeful behaviour. And if scientists working in systems analysis had always checked the techniques they just developed by the yardstick of explanation of, for instance, their own purposeful behaviour, I am quite sure that many of the already published papers would have never been seen in print.

Constant reference to human purposeful behaviour is compelled by a whole range of reasons.

First of all, systems analysis is inherently action-oriented, i.e. oriented towards our interaction with the object of study. To be able to furnish this orientation, we need to understand the laws of purposeful interaction and to comprehend what information is required by human beings to ensure the purposefulness of their actions, and what limitations are immanent.

Second, objects of systems study more often than not fall into the category of "human" objects. Hence, without an insight into the regularities of the purposeful behaviour of individuals and groups of people which constitute the object, we are unable to study it.

However, most relevant here is that the phenomenon of purposeful activity, by itself, should be the basic object of systems research. One is bound to agree that in purposeful activity, a human being constantly organizes and uses information about the objects of the very type we termed systemic. The environment of human activity is no doubt too complicated to be represented in the framework of a single model. This is only too well known to the experimental psychologists who indulge in "super-human" efforts to create experimental conditions which would force all the subjects to "use" one and the same model. Every interpretation of external events always contains factors of a diversified nature. There is always a need for interaction with other purposeful components involved in the action, and the problems facing a human being usually are ill-structured. As a result, human experience has already accumulated quite a few techniques to check the excessive complexity of the objects dealt with. In looking for a scientific foundation for systems research, it would be only natural to proceed from the study and further perfection of such "commonsense" techniques, methods and devices, established over many many years of everyday practice. In this case it is of primary importance not to lose sight of the initial object of study, the human being, and not to treat the concepts of purposefulness and other related concepts in an absolutized manner, but rather as a manifestation of certain properties of a specific biological object (conditioned by its potential and limitations).

It should be clear from the ideas expressed above that I am willing to regard the problem of study of the research development stage associated with the ideas of systems approach and systems analysis falling very naturally into the category of systemic problems. My opinion, as expressed in this chapter, should be taken only as one of the possible views of the systems analysis explanation, and in full compliance with the spirit of this conception, the existence of different complementary and at the same time partially contradictory interpretations is not only rejected, but should be regarded as an essential element of further development. That is why I hope that this chapter will be understood primarily as an appeal for a re-thinking of current developments and for the elaboration of other programmes that are of an integrated nature.

REFERENCES

1. W. R. Ashby, *An Introduction to Cybernetics*, London, Chapman and Hall (1956).
2. R. M. Axelrod, *Framework for a General Theory of Cognition and Choice;* Berkeley, University of California Press (1972).
3. I. V. Blauberg and E. G. Yudin, *Emergence and Essence of Systems Approach*. Moscow, Nauka (1973), in Russian.
4. G. M. Gvishiani, Materialism as a philosophical basis of systems studies, this book.
5. R. Mattessich, *Instrumental Reasoning and Systems Methodology*. Dordrecht, Reidel (1978).
6. E. L. Nappelbaum, Specific uncertainty facets in systems analysis problems, in: *Systems Analysis and Scientific Development Management* (Abstracted Proceedings of the Theoretical Conference). Moscow, Obninsk (1978), in Russian.
7. *Representation and Understanding: Studies in Cognitive Science*. New York (1975).
8. V. N. Sadovsky, *Foundations of a General Systems Theory*. Moscow, Nauka (1974), in Russian.
9. H. A. Simon, *The Sciences of the Artificial*, Cambridge, MIT Press (1969).
10. A. I. Uemov, *Systems Approach and General Systems Theory*. Moscow, Mysl' (1978), in Russian.
11. S. V. Yemelyanov and E. L. Nappelbaum, Individual and collective purposefulness in management, in: *Control and Management Problems*. Moscow, Control and Management Problems Institute Press (1975), in Russian.
12. S. V. Yemelyanov and E. L. Nappelbaum, The basic principles of systems analysis, in: *Scientific Management of Socialist Industry*. Moscow, Ekonomika (1974), in Russian.

SYSTEM AND WHOLENESS CONCEPTS

I. V. Blauberg

In systems research the concept of the whole, wholeness, plays an important role. Most authors base their studies on the fact that a system represents an integral set of elements. At the same time, there is a century-old history of the philosophical categories of part and whole, during which content, as well as methodological significance for the development of scientific knowledge, has been substantially transformed. It seems that methodological difficulties in cognizing and constructing complex integral objects, which concrete scientific and technological knowledge, experienced in the second half of the twentieth century in large measure necessitated the systems approach.

Considered as philosophical categories, the part and the whole express the relation between a set of objects and a connection which unifies these objects and is responsible for the appearance of the new (integrative) properties and regularities that are not present in isolated objects. The type of connection between the parts also determines the type of the whole formed.

The categories of part and whole also characterize the general motion of knowledge; this usually begins from the undivided concept of whole, to be followed subsequently by the analysis and the decomposition of the whole into parts, and is concluded by the reproduction of the object by thinking in the form of a concrete whole. Marx formulated these laws of cognition of integral objects in his Economic Manuscripts of 1857-1858. In this connection, the nature of the interpretation of "part" and "whole" categories, as well as the problem of wholeness derived from them, determine to a considerable extent the general strategy of scientific cognition, the method of solving basic scientific problems.

Here we are faced with the necessity of distinguishing between two meanings of the concept of wholeness. The concept of wholeness in a narrow sense is associated with the cognition of certain real objects characterized by complex organization (this is the meaning involved, for instance, in the problem of wholeness in biology, psychology, etc.) and with the general specific features of these objects. Wholeness in a broad sense is correlated not with one concrete object or another as such, but with a style of thinking typical of a given epoch; therefore, the interpretation of this concept is related to a certain kind of philosophical and general scientific reflection and creates a particular general background against which the motion of philosophical and scientific knowledge displays itself.

In view of the fundamental significance of the problem of wholeness for the theory and practice of modern systems studies, the question naturally arises about the comparison of the concepts of wholeness and of the systems property (the whole and the system). Anyone coming across the use of these concepts in scientific literature will surely have noticed that their contents are very closely related and that there is also a profound internal relationship between them. But what is the degree of this relationship: are they synonyms or are they connected by a different relation? To put it differently, what is the meaning of the use of these two concepts in scientific knowledge?

There is no convincing (or at least sufficiently detailed) answer to this question in modern literature which analyses the philosophical aspects of the problem of wholeness [1]. The appearance of publications where it seemed impossible to evade this question (our allusion is to literature completely or partly devoted to the definition of the system concept), as well as the development of other purely systems studies, has not done much to clarify the relation between the systems property and wholeness.

V. Sadovsky has devoted several publications to a typological analysis of diversified meanings of the system concept conducted on the basis of approximately forty definitions of the concept that were most common in monographs and papers on systems research. Mainly using his results, I will consider the forms in which the relation between concepts constituting the subject-matter of this paper appears in systems literature.

In ascertaining the fundamental diversity of reasons why the properties of a system are ascribed to certain objects, Sadovsky emphasizes the extreme complexity of the problem of determining the lower boundary of the systems property; at present there are no criteria for making the upper boundary of the systems nature more precise either; crossing this boundary means entering the field of various kinds of systems, rather than systems in general. Hence the problem of constructing a certain hierarchy of system properties, whose lower members would apparently be regarded as non-systems, and whose higher ones might possibly cover certain types of systems; on the whole, however, this hierarchy would specify the properties of a system as such [2].

Following this path, Sadovsky identified several informal features (considering their "systems" nature beyond any doubt) and divided them into three groups denoted by A, B and C. Group A included features characterizing the internal structure of a system: set, element, relation, property, connection, interaction, subsystem, organization, structure, etc. Group B contained features characterizing the specific systems properties: isolation, interaction, integration, differentiation, centralization, decentralization, wholeness, stability, feedback, equilibrium, control, and so on. Group C included features pertaining to system behaviour: environment, the state of a system, activity, wholeness, functioning, variation, homeostasis, purposefulness, etc.

Note that in this series of systems features that of wholeness appears both in group B and group C. Thus, wholeness is considered not only as an essential systems property but also as a characteristic of system behaviour. As far as group A is concerned, one can easily see that all or at least most of the features relating to it serve to describe the system as a definite whole.

Based on his consideration of the definitions of the system concept, Sadovsky concludes that they can also be divided into three different groups. The group whose volume is the most substantial and which is most interesting from the methodological point of view includes definitions the system through such concepts as "elements", "relations", "connections", "whole" and "wholeness", although the terms establishing these concepts in one definition or another are extremely heterogeneous. Sadovsky infers from this that the definition of a system through elements, connections, relations and wholeness forms the base structure of the definition of the concept, which describes at least a large class of systems, if not all system formations.

As for the two other groups, one covers the definitions that give concrete expression to the base definition through the introduction of additional attributes, and thus specify certain classes of systems (cybernetic, biological). I consider the second group, which includes definitions of a system as a certain class of mathematical models, to be a mathematical expression of the base definition (if the objects are described that satisfy this definition) and also as a broader class of mathematical models in terms of which one can trace the gradual transition from the non-system to the system object of study [3].

The relation between wholeness and systems analysed here forms a more generalized criterion of the classification of the definitions of the system concept. In accordance with this criterion, the set of these definitions is divided into two groups, one of which includes wholeness as an essential feature of every system, while the other contains no such feature. The first group fully covers definitions possessing the base structure (in terms of Sadovsky's classification) as well as their additional concretization, and also some definitions formulated in terms of mathematical models, namely, those which serve to describe objects with wholeness properties.

In the definitions of the second group, the system concept is usually interpreted from the set-theoretical point of view: the system is regarded as a set of elements with relations defined for this set. A "classical" example is the frequently quoted definition of A. Hall and R. Fagen: "A system is a set of objects together with relationships between the objects and between their attributes." Although it is exactly the assumption that the system has properties, functions or goals which differ from the properties, functions or goals of its components [4] that the authors base their argument on, one can easily see that this assumption is not explicitly established in their definition.

However, as has been demonstrated in several articles by Soviet scientists, interpreting the system concept in terms of the set theory [5] is not adequate for describing specific systems formations and should be regarded only as auxiliary analytical means of analysis. Yu. Shreider underlines the following fundamental difference between the set and the system: when a set is formed, the basic components are the elements, certain combinations of which form one or another set; for a system, the feature of wholeness, i.e. the fact that it constitutes a certain whole composed of interacting (connected) parts, is primary. For a system the elements are not specified beforehand; they are constructed (or selected) during the system's segmentation, each system permitting various forms of segmentation. Although each segmentation is a set, the system itself is not a set; it can only be regarded as a set [6].

If, on the other hand, the primary attribute of a system is wholeness, this cannot be described in terms of relations either, since the logical characteristics of binary relations (or those reducible to binary) do not allow one in principle to express the properties of integral organization (a special analysis of this question can be found in A. Anglyal's book *Logic of Systems* [7].

Thus, the second group of system definitions does not define this concept and can be ignored. A consequence of this is that any definition of a system which can be regarded as equivalent to its object includes the feature of wholeness as the most essential and decisive attribute of any system.

Although, when compared to the concept of wholeness, the system concept appears to be far more detailed and separated into elements, the feature of wholeness is usually accepted in the definition and description of a system as intuitively obvious and is not subject to further explication. In most cases, the wholeness of a system is specified through the indication of the mutual connection between elements; in his concept of "the general system theory", L. von Bertalanffy, for instance, for whom the study of the system was equivalent to that of wholeness, characterized wholeness as a property in the presence of which the variation of any element affects all other elements and leads to the variation of the whole system, and conversely, the variation of any element depends on all other elements of the system [8]. Without at all underestimating the importance of the mutual connection and interaction of parts during any characterization of the whole, we would still like to emphasize that the main element of this characterization is the property of integrativity, i.e. the appearance, as a result of interaction between parts, of new qualities and properties at the level of the whole, of new qualities and properties which were not inherent in the individual parts or their sum.

It is exactly the integrativity of the whole, and not the interconnection between the parts in itself, which provides a clue to the explanation of the fundamental role of the concept of wholeness in scientific knowledge. Without establishing this aspect of the question, our position would hardly be different from that of the eighteenth-century French materialists, for whom the principle of interaction between the parts of a whole was indisputable.

It is also important to note that in systems research literature, the relation between the concept of the system and that of the whole is usually not made more precise either, with the result that these concepts turn out to be interchangeable within quite broad limits. If one compares the interpretation of the concept of the whole in publications on the wholeness question to that of the system concept in systems literature, one can see that both concepts are in principle characterized by the same attributes. The whole, as well as the system, consists of parts (elements); the stable, invariant connections between parts form the structure of the whole; the whole, like the system, possesses the properties of organization and orderliness. The whole is hierarchically organized *ex definitione*: what is a part in some respect is a whole in another; therefore, every whole is a part of another whole, and any part is in its turn a whole. In the analysis of the whole, just as in the study of systems, the insufficiency of purely analytical segmentation of an object is revealed, the necessity of using synthetic research procedures comes to light, and so on and so forth. From the similarity of those basic specific features of the system and the whole, one arrives at the similarity of other less essential peculiarities, so this enumeration could be easily continued.

The following conclusions can be made:

(1) The great majority of those writing on systems problems do not distinguish between wholeness and the systems concept (between the whole and the system). If in definitions of the system concept the attribute of wholeness is not present in either its explicit or implicit form, these definitions are not definitions of the system proper.

(2) Some of those writing on the problem of wholeness believe that the concepts of the whole and system are essentially identical.

Authors distinguishing between the concepts refer to one or another definition of the system (usually to that of Bertalanffy: the system represents "a complex of elements standing in interaction") to confirm their point of view. But references to a certain system definition which is universally recognized in scientific literature are not convincing, since the extensive development of systems studies has recently been accompanied by the appearance of such a great diversity of those definitions, both informal and formal (discussed above), that an attempt to identify certain uniform contents in them would require a considerable effort of classification.

Thus, it seems to us that returning to the question of the "degree of kinship" between the concepts of wholeness and the systems formulated at the beginning of this paper, we should answer it in the following way: yes, these concepts express the same content and are essentially synonyms. Therefore, on the basis of purely stylistic considerations, they can easily be substituted for each other in corresponding texts, or the gradual disappearance of one of them in special scientific literature (most probably, the concept of the whole would die out, as it is less "modern") is just a question of time.

This answer, however, is not satisfactory.

As we perceive things, the key to the development of science is a principle which goes back to Occam and amounts to the statement: we must not invoke more kinds of entity than we need for an explanation. For all their profound kinship, the concepts of the whole and the system, with a history of over a century and continuously reproduced in the evolution of modern scientific knowledge, cannot help possessing substantial differences. To identify the specificity of each, however, we must leave the narrow circle of definitions and competing interpretations of these concepts and consider them in a broader context, namely, the context of the functions they perform in modern science.

Thus, we should now answer the second half of the question formulated at the beginning of the paper: what is the meaning of the functioning of these two concepts in scientific knowledge? This is exactly what we are going to do now, being aware that the attempt, as it is the first of its kind, may turn out to be unsuccessful; at any rate, we hope that it will be helpful for further discussion of the issue.

I will begin with difficulties which are encountered by anyone who tries to record the wholeness criteria (and, accordingly, those of the system) from the point of view of a purely ontological approach to various objects of reality. A recording of this kind implies that it is possible to formulate a certain (sufficiently complete) list of features typical of all integral objects, and of them alone. In other words, a certain class of objects should be singled out at the empirical level, a class which possesses the above features of wholeness, and differs fundamentally from another class of (non-integral) objects, which does not possess such features.

Above I tried to demonstrate how difficult it is to outline an empirical boundary separating the integral formations from non-integral formations. Now I would like to add that the problems that emerge in this instance are associated not with how apt a definition of the whole in a certain publication may be, but with the fact that the meaning of the wholeness concept can never be reduced to the function of generalization with respect to the level of scientific knowledge achieved at any specific point in time.

One should not underestimate the significance of this generalizing role of the wholeness concept, which is specifically expressed in definitions describing the general properties of integral formations. More precisely, each of these definitions records the general element that has been singled out in various objects appearing as whole at each stage of scientific development. But one must clearly be aware of what cognitive procedures these definitions are based on, and of exactly what functions they can actually perform in cognition.

When definitions of wholeness are constructed, the "upward" way associated with the initial recording of the features corresponding to the lower type of wholeness, followed by the subsequent addition to this set of features characterizing wholeness formations of higher types, is certainly inappropriate. For this kind of approach the resulting set of features depends essentially on the type of wholeness chosen as initial, but it is exactly this question which cannot be solved at the empirical level: the answer to it is related to the presence of a certain theoretical concept of wholeness, i.e. it implies that the definition of wholeness already exists in knowledge.

The "downward" way, i.e. the attempt to extrapolate the properties of highly organized types of wholeness so that they include all other types of integral objects, is as unproductive as the other. First, the same problem of defining the "lower boundary" of wholeness crops up. Second, the concepts of wholeness corresponding to the higher level of organization, developed by science, cannot be regarded as a certain accepted formula that would determine the understanding of other types and levels of wholeness. There are specific properties inherent in each of these levels, and understanding them implies the development of special concepts of wholeness of a given type and level.

It may initially seem that the ontological status of the system concept is somewhat different from that of the concept of wholeness, since its connection with its empirical "assessors" is less rigid. In fact, in systems literature the interpretation of the system as an arbitrary set of objects with relations (connections) specified for it is rather widespread. As an example, the system definition suggested by L. Blumenfeld at a discussion on a systems approach to biology (1968) can be considered: "The concept 'system' denotes a set of real or imaginary elements chosen in any way from the rest of the world. This set represents a system if (1) the connections between these elements are specified; (2) each element is considered indivisible inside the system; (3) the system interacts as a whole with the external world; (4) the set will be regarded as a single system if a one-to-one correspondence can be established between its elements at different moments of time; this statement applies to the case of evolution in time" [9].

As a matter of fact, however, if one is thinking of real systems research and not a certain logical mental game this arbitrariness in specifying the system is only apparent. In publications on the subject, one can frequently come across W. Ashby's statement about the possibility of imagining a system with the following three variables: the ambient temperature in a given room, air humidity, and the dollar rate of exchange in Singapore [10]. Incidentally, he himself admitted that this choice of variables is irrational and hardly anyone would disagree with him. In the above definition of the system concept, as well as in several other definitions, the importance of the epistemological position in the study of a complex object is correctly emphasized. Even so, it is quite obvious that this operation as specifying the connection between elements appears in reality as secondary, derivative with respect to the segmentation of a given complex object forming a certain wholeness (or is regarded as a whole). It is exactly this segmentation which can be performed on different grounds that makes it possible to identify certain elements of the system being studied.

It would be appropriate to repeat here that a systems study, unlike the analytical approach, implies motion not from the parts to the whole, but from the whole to the parts. This statement has to be brought to mind more than once and specifically it was mentioned previously, in a section where Shreider's point of view on the inadequacy of the set-theoretic language for the description of systems formations was presented. As I see it his reasoning is an apt concretization of a statement concerning the primacy of the integral approach over the elementarist one (based on elements, parts), and the possibility of diversified segmentation of an integral object, which occurs persistently in systems literature. In this connection, I would like to say that Sadovsky's classification of the components of those system concept definitions which are most common in the literature [11] seems to be made somewhat more precise. Among these components Sadovsky singles out the characteristic of the initial formations (A_1) - elements, parts, etc. and the characteristic of a combination of such formations (A_2) - complex, set, and so on. It is unnecessary, from the point of view of the methodological principles of systems research to provide special proof of the fact that it is A_2, i.e. the whole, and not the elements, which should be the primary item in the above enumeration.

In this case, therefore, one again discovers that the content of the system concept is inseparably connected with that of the concept of wholeness; for this reason it is just as difficult to "tie" it as it is to "tie" the latter to a strictly definite class of the objects of reality in order to characterize it exhaustively on the basis of the features of exactly that class of objects. In view of this conclusion it is natural to try and apply another method to establish the relation between these two concepts: to compare them methodologically, bearing in mind their meaning, place and functions in scientific knowledge.

The methodological functions of the concept of wholeness were identified and described by B. Yudin, and in our joint publication [12]. In the above, along with the generalizing definitions of the wholeness concept, other kinds of definitions were considered which, instead of recording the level of knowledge achieved (this is a function of generalizing definitions), appear as guidelines denoting the direction of further development of scientific thinking. Such definitions may be called methodological, since they cover not the objects themselves but the cognitive situations which arise in the course of the study of these objects. Moreover, methodological analysis helps reveal more effective ways of studying "whole objects" selected from among others used by science. Such analysis creates avenues of research which may first be applied to a narrow, strictly defined class of objects. This means that methodology, far from limiting the research possibilities of sciences, stimulates their development.

One of the substantial characteristics of such cognitive situations which reveal the methodological functions of the concept of wholeness is their ability to single out the initial element (structure) of analysis and to outline a theoretical interpretation of a certain sphere of reality. This "structure" acts as an element which is indivisible within the framework of a given level of research, and for this reason is inevitably viewed as one whole. Such understanding of wholeness is always inherent in any research, although it may not always be understood as such by the researcher himself. But even if at the start of the research this idea is more implied than clearly stated, it may nevertheless create a sound basis for streamlining the cognitive apparatus and for systematising the existing knowledge of a given object. The knowledge of the sphere of reality under study becomes part of systematic theoretical knowledge only after it has been interpreted from the point of view of this one integrated whole; and this enables us to make conscious use of a large part of the cognitive apparatus and particularly of such concepts as behaviour, functioning, development, structure, organization, etc. For example, it is hardly possible to distinguish between the process of functioning and the process of development outside the scope of this "one whole" which both functions and develops. We regard one and the same process either in functioning or in development, keeping in mind the existence of different "wholes". Thus, the process of development of an individual organism (ontogenesis) at the same time functions in another "whole" – a population or a species.

The above concept of wholeness serves as a starting point for a new coordinate system of theoretical explanation. We do not mean the cognition of the specific characteristics of an object as such, because here our "whole object" acts as the elementary structure of our interpretation, but not as the object of it. The situation is different when the cognitive process centres on the "whole object". (The term "whole object" is interpreted here as an object which cannot be understood if we proceed only from exterior elements.) This proves that in this case purely reductive means of interpretation are ineffective, since the objective of our research is to reveal the inner patterns governing the qualitative features of the object under study. The initial cognitive situation suggests that we already know something about this object. However, this knowledge is insufficient to fully understand all the particulars of the given whole. Hence a gap between what has already been understood and what is yet to be learned as registered by our comprehension of the object under study as one integral whole. This gap, however, is rather relative as we are clearly aware of what exactly we do not know, i.e. the cognitive process is directed at the interrelationship which exists in reality, but which has not yet been fully revealed. This clearly outlined part of reality which we do not yet understand makes the examination of an object a well-organized process. It follows certain specific laws which also have their inner logic.
What then, is the emergence of this function of the wholeness concept in scientific knowledge associated with?

To answer this question, we will require the method of distinguishing between the two meanings of the concept of wholeness introduced at the beginning. Here we are alluding to wholeness in the broad sense of the word, as associated not with concrete objects of complex organization but with a style of thinking characteristic of a certain epoch (culture). Now, when this is taken into account, it becomes clear that this categorical formation possesses a rather complex structure. Without attempting to provide a detailed description, which is as yet difficult, we still can single out two layers and append the conditional names "actual" and "potential".

The first layer consists of the concepts of wholeness, existing in philosophical and scientific knowledge at any given moment of time, and considered in their general form (the relation between the whole and the parts, the place of these categories among others, the interpretation of the system of concepts associated with wholeness, the concepts of the methods of cognition of the whole, etc.). It is exactly at the level of this layer that the generalizing definitions of the wholeness concept are located, although not completely, since they are also related, on the other hand, to the interpretation of wholeness in the narrow sense of the word.

The second layer constitutes that which is lacking in the previously existing concepts of wholeness; what is being found out about wholeness during the subsequent development of science; those relations that exist in reality but have not yet been discovered by scientific knowledge (although it is important for us that these are not the specific features of individual objects as such, but the specific features of the cognition of wholeness in general which are concrete in their application to individual objects).

It is quite obvious that this second, "potential" layer can be explained most completely in a historical scientific study when the evolution from one stage to another in the cognition of a certain integral object is traced and the differences between each subsequent stage and the previous one are recorded. But it would be wrong to assume that this layer manifests itself only in retrospect. It is, as it were, invisible though tangible, it is revealed in actual knowledge, but as an outline, a hint, a fragment, and not as a whole. The origins of these fragments or hints are easily understood: the whole, which is yet unknown to us, cannot help being in some way revealed in parts that we already know; the unknown structural level cannot help being somehow discovered at the levels already known, etc.

Let us venture the assumption that it is exactly the presence of this "invisible" layer which constitutes the source of intuition, creative thinking, etc.

In this connection, I would like to present an analogy from a different sphere. The well-known Soviet writer K. Paustovsky, reflecting in his *Golden Rose* on the role of details in a work of literature and the principles of their selection by the author, wrote: "The detail is most closely related to that phenomenon which we call intuition.

I conceive intuition as an ability to reproduce the picture of the whole from an individual detail, from a certain single property.

Intuition helps historical writers reconstruct not only the true picture of life of past epochs, but their very spirit, the very state of the people, their mentality which was, of course, somewhat different than ours...

And for the reader, a good detail produces an intuitive and correct conception of the whole: either of a person and his (her) state, or of an event or, finally, of an epoch." [13].

One can say with certainty that science is aware of many facts that represent fragments of a whole which is yet vague and which will be discovered later in intuitive inspiration. Perhaps it is precisely the knowledge of how to approach these facts (and they most frequently "fall out" from the group of generally recognized concepts and systems of thought) as a manifestation of some unknown level or unknown whole that is one of the distinctions between creative and non-creative thinking.

Thus, the introduction of a distinction between two meanings of the wholeness concept demonstrates that heuristic significance in scientific knowledge above all applies to the concept of wholeness in the broad sense of the word, with its "potential" layer. It is exactly this concept that forms the guideline of scientific knowledge, promoting the explanation of new facts or phenomenon from the view of a broader whole. And the concept of wholeness in the narrow sense of the word (associated, as we have already said, with cognition of objects with complex organization) is directed primarily towards the recording of the fact that a given concrete object of scientific cognition is integral, and therefore, the application of purely reductionist methods of study in the final analysis turns out to be ineffective.

Now we are in a position to make the relation between concepts "whole" and "wholeness" more precise. The "whole" is the result of applying "wholeness" as a concept that performs methodological functions to a certain concrete object, which is regarded from this moment onward as a whole provided it satisfies the criteria we adopt. It is hardly necessary in the present context to justify in detail the fact that this procedure is not at all a formal statement of a generally known fact, but is associated in most cases with the fundamental restructuring of the existing concepts of the object being studied, it is sufficient to recall the substantial shifts in linguistics as a result of the concept of language as an integral formation or the changes produced in biology by the similar concept of biogenocenosis (ecosystem).

Thus, the concept of wholeness in the narrow sense of the word generally coincides with that of the whole. "Wholeness" proper, however, coincides with the concept of wholeness in the broad sense of the word. It is with this meaning implied that these concepts will be used in the further discussion.

What is easily apparent is that the concept of wholeness in our interpretation cannot be described in formal language in principle, since it records, as was shown above, not so much and not only actual knowledge as the incompleteness of this knowledge.

What is not yet known, cannot be formalized. Thus, the frequent call in systems literature for an explication of the wholeness concept may be referred, in reality, only to the whole or more precisely, to the ways of expressing it in knowledge. It is the system concept, as well as the whole set of conceptual means associated with it, that plays the role of an explicate of this concept (whole). Therefore, the concept of the system is indeed inseparably connected with that of the whole.

So when we speak of the relation between the concepts "whole" and "system" we in fact admit a certain simplification. Actually, the point is that a certain integral object and its properties are described by means of a developed set of concepts, whose central and organizing place is the concept of system. A complete and accurate description of an integral object through systems conceptual means presupposes that the given object has already been singled out in knowledge as a certain integral formation and that the researcher has to explicate the connections and the properties of this object and, as far as possible, to express them in the corresponding formal-mathematical concepts and operation procedures. To put it differently, the system concept includes those aspects of the study of an object with complex organization which have already been "worked out" in the process of using the concept of wholeness and which admit of formalization. What this means is that the methodological role in systems research is played not by the concept of the system but by that of wholeness. And in this case systems studies themselves appear as an important "proving ground" on which the methodological-heuristic function of the wholeness concept is "tested".

It follows clearly that the increasingly extensive use of the system concept in modern scientific knowledge does not at all imply rejection of the concept of wholeness, since it is this latter which specifies the methodological context of scientific knowledge in which systems studies evolve. At the same time, the concept of wholeness does not allow for explication; therefore, to characterize it rather nebular expressions like the style of thinking", "the background of scientific knowledge" and "the type of reflection" have to be used. We try to describe the concepts of whole, wholeness, system and their interrelations more specifically.

The whole (*das Ganze, le tout*) represents a concrete object with integrative ("emergent") properties. From the epistemological point of view, integrativity appears as the resulting product of the generalizing function of the concept of wholeness, associated with the cognized features of objects with complex organization.

Wholeness (*die Ganzheit, l'intégrité*) represents concepts of the completeness of the coverage of phenomena, and at the same time, of the essence of integration, the processes of formation of new, structural levels, hierarchical organization of processes and phenomena, etc. which exist at every given instant in philosophical and scientific knowledge. This is the background against which the cognition of integral objects and the guideline of cognitive activity develop. The concept of wholeness performs this function because it has a peculiar two-layer structure which includes not only actual but potential knowledge.

Hence it is clear how the "whole" relates to "wholeness". The whole is a concrete object (a class of objects) in which, through the application of the corresponding research procedures, the presence of integrative properties is discovered. Thus, the concept of the whole is formulated as the result of applying the concept of wholeness, and is associated with the realization of cognitive activity rather than being the primordial characteristic of the object by itself. Therefore, the development in any science of adequate concepts of the object being studied as a whole is a very important stage in its evolution.

The system is a concept which serves to represent in knowledge an integral object by means of specific principles, certain conceptual and formal means; this representation is usually undertaken with a definite practical objective (for instance, in connection with control problems). It is recognized that the representation of an integral object (a whole) as a system is not the only possible form of its reflection in knowledge, as there may be structural, functional, structural-functional, elementwise and other representations. One should not exclude the possible appearance in this representation of different and more efficient methods of the description of the whole, which might replace the systems approach in the same way as it has replaced others.

What is the relation between the system and wholeness concepts? As follows from the above, the system concept always describes the whole and is inseparably connected with it (thereby, it is also connected with the concept of wholeness). Wholeness, on the other hand, is not exhausted by systems description because of the impossibility of formalizing this concept.

In the real process of scientific cognition, the concepts of the whole, wholeness and system form a certain hierarchy that includes, in addition to the above concepts, several others which are related to them rather than occupying adjacent places. This integral hierarchically organized conceptual system represents a subsystem of scientific knowledge as a whole, considered from a definite angle. The concept of wholeness appears here as an integral description of the synthetic tendencies of scientific knowledge.

It is exactly with respect to this type of system of concepts that the place and the significance of each individual concept (the whole, system, structure, organization, etc.) and their hierarchical subordination are determined in it. Obviously, these concepts are not synonyms, and each of them expresses quite definite content. At the same time, it is equally impossible to draw the line of absolute distinction between them from the point of view of their content and meaning because of their close interrelation in the system.

To sum up, one can conclude that the study of the methodological aspects of the problem of wholeness and the development of the systems approach can only be realized under present conditions in their inseparable connection with each other.

REFERENCES AND NOTES

1. For an analysis of the literature on the problem of wholeness in the light of the above relation between concepts, see I. V. Blauberg. Wholeness and the systems concept, pp. 6-15, *Systems Research Yearbook 1977*. Moscow (1977), in Russian.
2. For a detailed treatment of this topic, see V. N. Sadovsky, Some fundamental problems of constructing the general systems theory, pp. 47-48, *Systems Research Yearbook 1971*. Moscow (1972), in Russian.
3. Ibid., pp. 51-53.
4. A. D. Hall and R. E. Fagen, Definition of system, *General Systems* 1, 18 (1956).
5. E. R. Rannap, Systems analysis of invention description, *Scientific and Technical Information* Series 2, No. 6 (1971); Yu. A. Shreider, On the definition of system, *Scientific and Technical Information* Series 2, No. 7 (1971); V. N. Sadovsky, General systems theory as a metatheory, *Voprosy filosofii* No. 4 (1972).
6. Yu. A. Shreider, Op. cit., p. 5.
7. A. Angyal, *Logic of Systems*. Systems Thinking, New York (1970).
8. L. von Bertalanffy, An outline of general system theory, *British Journal for the Philosophy of Science* 1 (1960).
9. *Systems Research Yearbook 1970*, p. 37. Moscow (1970), in Russian.
10. *Studies in General Systems Theory*, p. 129, (a collection of translated papers). Moscow (1969).
11. V. N. Sadovsky, Some fundamental problems of constructing general systems theory, *Systems Research Yearbook 1971*, pp. 51-52. Moscow (1972), in Russian.

12. B. G. Yudin, The concept of wholeness in the structure of scientific knowledge, *Voprosy filosofii* No. 12 (1970); I. V. Blauberg and B. G. Yudin, *The Concept of Wholeness and Its Role in Scientific Knowledge*. Moscow (1972), in Russian.
13. K. G. Paustovsky, *The Golden Rose*, Collected Works, Vol. 2, pp. 614-615. Moscow (1958), in Russian.

ACTIVITY AND SYSTEMS APPROACH
E. G. Yudin

The notion of activity is one of those concepts which does not need any special validation of its relevance. In modern knowledge, and especially the humanities, the concept of activity plays a key, methodologically central role, because it is through the concept of activity that human beings gain the most universal and fundamental explication of the world around them. Naturally, one of the manifestations of this concept is present in any methodological analysis and, to be more specific, there can be no single validation of any particular subject-matter of the humanities without it.

When the concept of activity is used with certain methodological implications, it usually means either the natural-historical basis of the life of a human being and society, or a reference to a special reality demanding specific research techniques. In both cases the content of the concept "activity" is specified and revealed through other concepts. In the former instance, the objective of the specification is to elaborate the ideas of the general structure of activity and its typology (similar, say, to the Hegelian scheme of activity that involves concepts such as goal, means and result as components; the distinction between material and ideal, productive and reproductive activity, etc. may be treated under the same category). In the latter instance the concept of activity is amplified through the development of the conception of its mechanisms which are presented in terms of the relevant specific discipline (psychological theory of action, verbal activity in linguistics, etc.). A very pertinent fact is that the two instances of the concept usage are intimately connected; the analysis of activity as a specific reality in fact implies the need to find an appropriate projection of a more general concept.

It is this close link which seems to engender a peculiar methodological phenomenon which has lately become increasingly prominent. At a certain point in the evolution of a specialized intradisciplinary conception of activity, one becomes aware of its partial and incomplete nature and its dependence on the general concept of activity. Of course, this was also understood in the past, since each of the humanities in the final analysis deals with a certain projection of activity (regardless of how clearly this fundamental methodological issue is understood). Hence the rethinking of any of its subjects, whatever the reasons, has always been related to the understanding of the dependence of that specific projection on a more general concept.

However, the process has acquired certain features at the present stage, and we will mainly be concerned with two. First, in methodological reflection, a more definite and explicit use is made of the category of activity, rather than of its substitutes, as was done before. Second, realization of the partial and incomplete nature of the specific projection of activity often inspires the researcher to look for the remedy in a systems approach. The latter phenomenon is most prominent in modern psychology, confirmed by research of authors as different in many respects as G. Allport [1] and A. N. Leontyev [3].

It is in this manner that the concept of activity merges with systems approach ideas. For modern thought, this merger is definitely in character - systemic concepts are used extensively in modern science, and at least nobody finds it surprising when confronted with their advent into still another field of research. But the problem is not all that simple when studied more thoroughly. Although recognition of the systemic nature of activity might seem attractive, this alone is not enough even to claim a first step in the elaboration of a new trend in the study of activity. The experience of various scientific disciplines demonstrates quite clearly that resorting to a systems approach is rational and effective only when it leads to the formation of a new object of investigation. With activity, the situation becomes additionally complicated because special analysis is required to identify relevant characteristics of the newly-formed discipline, its type and, more specifically, its relations with other disciplines within which the activity has been studied.

So the merging of activity studies with a systems approach in its original form only helps establish the rather trivial fact that activity is an extremely complicated research subject, and that within the framework of existing disciplines, the complexity cannot be represented adequately. But what are the criteria of adequacy in this respect? Because in this particular case our primary subject of study is activity, the criteria obviously have to be dictated by the concept of activity itself, although not by the systemic principles as they stand. To answer the question, therefore, it would help to analyse the concept of activity more thoroughly. To elucidate the topic of discussion we will assume as basic the definition of this concept from [5]: activity is a specifically human form of active attitude to the surrounding world, aimed at its purposeful modification and transformation by adopting and furthering available forms of culture.

This definition can hardly be regarded as exhaustive, but it does reflect one of the most fundamental features of activity, i.e. its universality, and, consequently, is a convenient point of departure in the analysis. True enough, the position and role of the concept of activity are determined first and foremost by its belonging to the class of universal, ultimate abstractions. Abstractions of this genre are the embodiment of an ubiquitous purport: they furnish meaningfulness to most elementary acts of existence and its most profound foundations, whose penetration makes the true integrity of the world intelligible.

The fact that these concepts may have direct reference to any elemental form of existence gives them an extremely solid ontological status, and to appeal to them is equivalent to evoking something totally self-evident and requiring no further definition. At the same time, they have an intuitively authentic content, although it is concealed from direct speculation and this relates them to the ultimately conceivable scope of existence and to its essential

explications, due to which the same abstractions can play the role of a
"final" argument, the most convincing and decisive testimony in validating
the speculative constructs of different nature and purpose. Although there
is the danger of oversimplifying the matter, we may say that abstractions
of this nature combine the empiric authenticity with theoretical profoundness
and methodological constructiveness. This accounts for their outstanding
role in the evolution of cognition: as they are rather limited in number they
(each time in some unique and specific combination) give an impression of
consolidating the space of speculation of the respective period, direct the
impetus of change in that space, and determine the type and the nature of the
objects of speculation in the period.

The universality of the concept type to which activity belongs is at the
bottom of still another of their features - multifunctionality, which is not
at all self-evident but crucial methodologically. Strange as it may seem,
the multifunctionality has long escaped the attention of scientists. A
possible explanation for this is the high "ontological" authenticity which
makes the concepts appear identical to themselves regardless of the context,
and thus they fulfil the same function. In fact, the function probably
cannot be the same. It is rather the opposite: it would be more natural to
assume that the universal ultimate nature of these abstractions should match
the multitude of functions they possess; specifically, this assumption is in
agreement with the essential versatility of ultimate situation types in
which they "work".

As for the concept of activity (i.e. when considering only connotations of
the concept pertinent to the context of scientific thought) at least five
different functions can be indicated:

(1) Activity as an explanatory principle - the concept with a philosophical-
methodological content conveying the universal basis (or, in a more careful
terminology, universal characteristic) of the human world.

(2) Activity as the subject of objective scientific study, i.e. something
dissectable and reproducible in the theoretical picture of a certain
scientific discipline in accordance with its methodology, specific tasks and
basic concepts.

(3) Activity as an object of control, as something to be organized into a
system that will function and/or evolve on the basis of a set of fixed
principles.

(4) Activity as an object of design, i.e. determination of the ways and
conditions for the optimal realization of certain (primarily new) forms of
activity.

(5) Activity as a value, i.e. the place that activity occupies in various
culture systems.

Even though the above list may not exhaust all functions of the ooncept of
activity, this is not the aim. The aim has been much more humble: to record
the real polyfunctionality of the concept of activity, and to show that in
various monodisciplinary constructions this plays an essentially different
role, and to show that one and the same word might often convey different
concept, To support this statement it is not necessary to consider all the
above listed functions of the concept of activity, but only to compare the
first two, i.e. activity as the explanatory principle and activity as the

subject of objective scientific study.

Among the things characterizing any system of theoretical thinking are the specific set of object-oriented inferential techniques and procedures, the most important of these being the concept and patterns of explanation, i.e. certain stable and rather stereotyped rules for the authentic acquisition of knowledge. The bases of these rules are a very small number of concepts with unconventional functions. Their unconventional nature is that in the process of acquiring knowledge they are backed not by other concepts (which is the case in a "normal" theoretical discourse: for instance, the concept of language is communicated via the concepts of culture, of speech and thought activity, of a symbolic system, etc., the concept of science - via the concepts of knowledge and its types, of social institute, etc.), but by reality itself. Thus it delineates a limit, a boundary, to the subject of thought. This is why concepts of this nature are sometimes called marginal abstractions (a basic and detailed analysis of their origin and functions is to be found in the book by M. K. Mamardashvili [4]).

But what is the meaning of this marginality? First, it should be emphasized that it has nothing to do with establishing the borderline between what has already been cognized and what has not. The main question in this case pertains to the organization of thought, its space and particularly, the specific idealizations used as the basis of all scientific thinking. Essentially, the purpose of any idealization is to provide a certain ultimate marginal situation - ultimately conceivable and thus exhaustive (in a logical space) for a certain class of situations. It is in an ultimate situation of this kind that scientific thought can, with exceptional accuracy and profoundness, bring to light the so-called laws of science: because to take the thought to the limits of idealization is to do away with the empiric diversity and create a theoretical object whose purpose is precisely to evoke the law-like conformity of the reality not inherent in the thinking. Hence, marginality in this case is a synonym of law-like conformity, the inner arrangement of the subject of contemplation, as well as the expression of its isolation, its detachment from other objects.

But I will not consider all and every idealization (of which, as is known, there are many in modern science), but only those marginal concepts which form a basis for the universal explanatory schemes and, consequently, can act as universal explanatory principles. Their marginality signifies that upon reaching the level of these concepts, thinking also reaches the limit in its contemplation of the object, after which it can penetrate deeper only by moving in the plane of the thought itself, rather than the object, i.e. through introspection, which under certain conditions may also open up new vistas of the reality but with references to a new and modified system of concepts. Their universality assumes that they characterize in the strict sense of the term a certain universe - the entire conceivable world conceptualized as an irreducible whole by reducing it to a common "scale" or a common dimensionality.

As is generally recognized, theoretical thinking has an established periodicity. Apart from all other things, every major period seems to tally with a distinctive inherent general explanatory scheme, within which one may additionally distinguish a certain universal concept serving as a logical epicentre. For instance, in classical thinking, the concept of cosmos played a universal role of this kind: it is to cosmos that the "final" explanations of reality referred. The science of the New Age became prominent parallel to the introduction in its thinking and meaningful development of the concept of nature to play the crucial role of the universal explanatory notion in contrast to the far more humble role allotted to the concept of φυσισ of the ancient Greeks or that of the concept of *natura* of the Romans. The new spirit of explication of the concept of nature was most forcefully voiced by Spinoza: his argument of nature of the *causa sui* asserts nature as the highest explanatory principle. In the framework of natural sciences, this meant that if a law or anything stable referring directly to nature were uncovered during investigation, it implied that the maximal conceivable limit of explanation had been reached.

The concept of nature is still a comprehensive explanatory principle in science; compared to the nineteenth century, let alone earlier times, its content has essentially been modified and expanded, and this explanatory potential of the concept of nature is still quite sufficient in solving most (if not all) fundamental problems in the sciences. In the broader sense, however, the comprehensiveness of the concept of nature as the explanatory principle was questioned even by Descartes: his proposition that extended and mental substances are equally significant essentially claimed that it was impossible to provide a purely naturalistic, scientific explanation to all the reality that man encounters. The need to introduce the concept of mental substance itself demonstrated the existence of a wide range of phenomena belonging to supranatural reality and thus requiring a special explanation. This in fact was the first allusion to the principle of activity as a substantiative and consequently, universal notion.

The next step was made by German classical philosophy. It not only identified activity as the "primordial substance" of the human world, but revealed the world as a genuine universe of activity. It is true, however, that Kant implies this more than he states it explicitly and one has to read between the lines to discern the principle of activity in Kantian apriorism and transcendentalism. But Fichte already builds his system as the philosophy of activism, and Hegel projects all human history on the activity principle and gives the concept - which is equally essential - a structurally developed expression through categories of purpose, means, and result. The latter is of the greatest possible importance, as it marks the transition from the empirically obvious fact of activity as captured in a superficial understanding of the term to the penetration of the structure and complex dynamics of activity, and its presentation as a reality dissected in a special way. Its presentation in this manner substantially broadened the explanatory potential of the concept of activity, and made it a true universal explanatory principle combining empirical authenticity with theoretical profundity and methodological constructiveness. But what has been gained specifically as a result?

First of all, a new explanatory principle has enabled perimeter of the reality which is the subject of social cognition to be traced. In any case, previous abstractions reduced the presentation of this reality to a set of individuals or psychological agents the mystery of whose existence is concealed in the brain of each. Meanwhile, activity placed in the centre of the universe dramatically alters both the framework and the entire fabric of the contemplation: the role of the primal point of departure now moves from the mere fact of the physical existence of these individuals to the presence of a profound meaningful interrelation between them. Taken diachronically, the relation presents the tableau of social world history and its synchronic aspect opens up a way to discover the principles of the functioning societal organism. In other words, the concept of activity essentially broadens the boundaries of the cognition of social reality and for the first time provides a theoretical basis for its presentation on a par with other historical processes of natural science.

Second, in these new confines social reality is represented as something with an inner organization conforming to certain rules. It is activity that acquires the status of both a source and mechanism of such organization, while the individual rests stripped of his initial sovereignty and emerges simply as the instrument of activity, only its element – however compulsory and irreplaceable this may be. Thus, the activity concept not only establishes the new boundaries for the reality, but also identifies the source of its inner conformity, and its natural modification, with the source itself being within the grasp of human cognition and devoid of any mysticism. This move has come as a much needed prerequisite to subordinating social cognition to the general canons of scientific theoretical thinking.

Third, the structurally dissected presentation of activity (which in fact has demonstrated how developed the concept is both theoretically and methodologically) can be used to explain and theoretically reproduce at least certain concrete phenomena of social reality. To support this statement one can allude to the Hegelian analysis of activity, which even at that time provided an essentially new vision of the subject-matter of linguistics, and opened new methodological vistas in study of the history of language and the development of historical comparative linguistics.

Important though it might have been, the Hegelian scheme proved far from perfect. Its main deficiency was the idealist nature of the basic premise Just as in all other classical German philosophy, the Hegelian substance is understood to be the spirit as some supernatural substratum, and activity is introduced into the analysis only as an attribute, albeit the most important one, of this substance – a manifestation of the spontaneous reasonable activity immanent to the spirit.

The postulate of reason had another important consequence: the category of activity viewed in this light was in fact reduced to the rational activity which fitted the "purpose-means-result" pattern. This rigorist rationalism was later to provoke the multifarious criticism of the Hegelian concept of activity. While dismissing the purely emotional impetuses and other aspects of that criticism, it is impossible not to retain at least one aspect which is of considerable methodological importance – i.e. some lines of critique employed as a point of departure the claim that activity was insufficient as an explanatory principle (referring to the interpretation of activity in the light of absolute rationalism) and tried to find the explanation and substantiation to the activity itself (either via the concept of will à la A. Schopenhauer and F. Nietzsche, through the concept of values as the neo-Kantians of the Baden school tried to do, or using the concept of unconscious or subconscious structures, as was the case with the Freudians and neo-Freudians). This certainly contributed to the understanding of activity, its nature and context, but at the same time it distracted from the conception of activity as the explanatory principle.

The concept of activity was further and decisively developed in Marxist philosophy and its socio-economic theory. While in German idealism the objective of ideological charge of the concept of activity was to reveal the active nature of the mind, in Marxism activity itself emerges as a true substance of culture, of the entire human world. Most naturally a change of this nature in outlook must go together with important alterations in interpreting the concept of "activity". A particularly significant point was the postulation of the object-oriented nature of human activity, introduced as early as "Theses on Feuerbach". This was a real revelation of the essence of the Marxist understanding of activity, i.e. its dual determination. In German idealism, activity was determined quite unambiguously: its only source was the immanent activity of the absolute. In contrast, Marxism postulates that activity is determined by both the logic of its object and its own inner logic. The first line of determination corresponds to the central thesis of materialism which gives priority to the objective and material, while the second expresses the creative and "open" nature of activity, its ability to create a "second nature". The introduction of the principle of object-orientation has led to the further extension of our ideas about the agent of activity, now treated not in terms of abstract idealism with the inevitable reverence of the psychological, but in objective-historical terms as the carrier of the totality of the dialectically organized activity of all society (first of all, material activity).

The role of the principle of activity within the framework of Marxism is generally known. From the general philosophical perspective, it makes it possible to build up an integrated materialist explanation of all social history and to reveal the internal structure of the society based on the division and transfer of activity. At the same time, the concretization of the principle in the realm of socio-economic knowledge is an important methodological condition for constructing a theoretical picture of the development of capitalism and its revolutionary replacement by socialist society.

The Marxist interpretation of the activity principle developed by Lenin exerted and still exerts a tremendous methodological influence on all realms of social cognition, and this influence has extended far beyond Marxist thought itself. In fact, the concept of object-oriented activity in one of its ramifications is employed in all the modern humanities. The explanation of social reality based on this concept enables us to cast aside the psychologism inherent in conventional humanism, discloses almost entire strata of a new, supraindividual reality, and promotes new and unexplored trends of analysis. Many different examples might be produced to substantiate these claims, but we will allude only to a couple of situations that are important typologically.

The first situation emerged in the modern study of science. Even the layman cannot fail to see that the scope of its research has broadened dramatically in the past few decades. Superficially, this is expressed in the fact that along with relatively conventional logico-methodological and pscychological analysis of science, sociological research has become equally prominent. But on a second look, we would be forced to admit that in itself the sociological approach to studies of science became possible only after the evolution of scientific knowlege was channelled to fit the explanatory scheme based on the principle of activity.

The point is not simply that along with the personality of the scientist and the product of his labour - scientific knowledge - a social institution with a developed system of activity was suddenly "discovered"; nor is the central point that we have just become aware of the decisive role of scientific activity as a factor in the generation of knowledge. The most important point in this instance is that the activity oriented explanation which penetrated this field literally forced us to take notice of the phenomena that had been obviously empirically established previously, but only now came to be recognized as the object of special investigation, as a specific product of scientific activity which in turn contributes to the understanding of the nature of activity. Apparently it is only in this context that communication in science might emerge as an object of special study: from the methodological viewpoint, to undertake such a step it was necessary not only to be convinced of the fruitfulness of the activity oriented approach in general, and one might say, accept the idea of activity, but also to conceive of the life of science as a complex system of activity with inherent specific accessories including communications, citation networks, etc.

Two aspects of this example are equally interesting. First, it shows that the concept of activity is not to be employed merely on verbal level and its application cannot be reduced to simple nominations. The concept of activity prompts a view of social reality which is able to cut out from the multitude of versatile overlays only those phenomena that fall into an integrated whole as the universe of activity, its products, conditions and forms of organization. Second, the example referring to the advent of new lines of inquiry into the development of science illustrates another previously mentioned point: one can easily see the way the explanatory scheme based on the concept of activity vitalizes new issues of investigation and gives a direct impetus to essentially expand the scope of research.

In this example, we encounter another curious phenomenon with a direct bearing on the topic: the integrating function of the concept of activity in studies of science is so obvious and suggestive that many scientists are tempted to use the concept as a basis for the development of an integral picture of the evolution of knowledge and science in general. Since it is known from the outset that activity and even its special type - scientific activity - are multidimensional and hence are unable to immediately fit the framework of linear presentation, systems concepts are suggested to design this picture. This engenders one of the specific versions of the merger between an activity oriented and systems approach. I will return to this point later, however.

The second topical example involves the development of the principle of activity in the history of Soviet psychology (to do it justice, one should note that the idea of activity-oriented explanation - though in a substantially different exposition - was rather fruitfully developed by representatives of the French psychological school beginning with P. Janet; but in this case we are concerned not with historical thoroughness but with a specific example of the conscious and systematic usage of the concept of activity as the explanatory principle). It is an established fact that L. S. Vygotsky, the founder of the cultural-historic conception in psychology, appreciated the enormous explanatory potential of the Marxist interpretation of the principle of activity and employed it to radically restructure the entire subject-matter of psychology. To demonstrate the core of his methodological achievements we will compare them with other similar efforts dating to approximately the same time.

In the 1920s there were a good many such attempts, all of which proceeded from the necessity to explain mentality as a special integrity. But the basis for this explanation was differently motivated in each case. By that time the conception of structural multilevel mentality (see [6]) had been widely accepted in psychology and there was a tendency to make any explanation - though not always explicitly - both structural and multilevel at the same time. For instance, according to the Freudian concept, the specifically psychological reality was to be found in the sphere of motivation [9] (which, in fact, is also one of special projections of the principle of activity) and the secret of the latter - at the level of the subconscious and its structural arrangement. Behaviourism, which concerned itself with the protection of methodological rigorism (in vain, as it soon turned out), discarded everything except for the behaviour parameters which could be recorded objectively; it was this reactive behaviour presented by the "stimulus-response" formula that was used as the explanatory basis and the integrating factor of the psychological reality in this conception. A special position case was Gestalt psychology, which remained indifferent to the idea of multilevel arrangement of mentality and shifted the focus in the theoretical explanation to the structural arrangement of psychological phenomena, maintaining the postulate of the priority of the psychical form.

Each of the schools, as well as many others, have developed a certain theoretical picture of reality to be studied by psychology. However, a closer examination shows that to attain the theoretical unity and logical homogeneity which are indispensable prerequisites of any subject claiming the status of a science, certain important inherently psychological features had to be dropped. Particularly appalling sacrifices were made in behaviourism, which was the reason it has never found its way into European psychology. As a result, the integrity of explanation which all theoretical constructions in fact aimed for has assumed a relative and essentially conditional nature, since the real integrity of mentality was reduced to a rather restricted integrity of one of its (and not always its) components. In other words, successful attempts to develop an explanation always entailed reductionism in one of its forms [11].

It should be emphasized that by itself the concept of activity is not safeguarded from reductionism. This is the reason why the methodological significance of the ideas of L. S. Vygotsky can not be reduced only to the introduction of the concept of object-oriented activity in psychology. Two basic circumstances aided the development of an efficient theoretical scheme based upon this concept.

First, it enables the reinterpretation of the whole range of mental manifestations or, at least, the entire set of higher mental functions through reference to object-oriented activity: the mentality was thus treated as a specific instrument or organ of activity, and this explanation, functional in nature (all other explanations did not have an expressed functional character), has added an important new dimension to the psychological interpretation of the integrity of psychological reality and its many specific manifestations. Hence in a purely methodological sense, the first achievement of the cultural-historic conception was the use of the concept of object-oriented activity as an instrument of the functional explanation and validation of the integrity of psychological subject-matter.

Second - and this is extremely relevant - the concept of object-oriented activity was "transplanted" into psychology not in its what one might call primordial form, but after the appropriate psychological interpretation. Vygotsky himself used the concept of interiorization as a main vehicle for the interpretation, whereas his followers, principally A. N. Leontyev, relied mainly on the concept of psychological structure of activity to do so, where the structural sequence "activity-action-operation" is parallelled with the "motive-objective-condition" sequence [3]. And this provided grounds for speaking of the psychological theory of activity.

But what was the specific explanatory function of the concept of object-oriented activity for that psychological school? First of all, it enabled mentality to be considered as a functional "organ" of activity, but in addition it opened the way to explaining the origin and development of the mentality: these two processes came as a direct product of the development of object-oriented activity. This conclusion, inferred directly from the activity scheme of explanation, is far more radical that it might initially appear. When applied to education, for example, it means that the most efficient way of forming the higher mental functions is to develop and to use the respective forms of object-oriented activity rather than simply acquiring knowledge which has in fact been the usual practice in the educational system. If school education is today the object of so much criticism, one of the main reasons is the failure to pursue the principle of activity in education [2].

So when used as an explanatory principle, activity can fulfil an important constructive function. However, the experience of specific scientific disciplines shows that this function is not self-actualizing; it is not predetermined by the concept of activity as such. The principle performs real and not illusory constructive functions only when it finds its object-oriented interpretation inherent in a field of knowledge under consideration. Otherwise, activity (as any other similar concept) serves only as an instrument of superficial attribution. Object-oriented interpretation assumes the observance of at least two important methodological conditions: first, the concept of activity should delineate a specific reality in its specific boundaries (reality *sui generis*, according to V. S. Shvyrev [7]); second, this concept must be structurally elaborated for a specific object of investigation as Hegel did for activity as a whole, or the cultural-historic school in psychology did for the problems of psychological analysis of activity. Structural development of the concept of activity encounters another very important problem: the necessity of an object-oriented interpretation of the concept of "activity" when used in a certain special field of knowledge signifies that even with an overall unit of the explanatory function of the concept, we in practice still have to deal with at least two of its "levels".

It is time that in the socio-philosophical analysis, the concept of activity is used to explain the entire conceivable human world with its incredible variety of manifestations; and to be able to attain this objective, one has to make the most of the idea of the boundaries of explainable reality and to imply a mechanism of its integrity. At the same time when the concept of activity is used as the basis of explaining a certain specific aspect of social reality, the general, socio-philosophical implications of the concept are evidently only a first step, to be followed by a second involving the elaboration of a problem area-oriented rather than general interpretation of the concept of activity. It appears that this implication of the concept cannot completely coincide with the former either in volume or in the meaningful characteristics of explainable reality and the mechanisms of its integrity.

The fundamental role which activity plays in the integration of social reality does not necessarily mean that reality in all its ramifications may be reduced to activity. There are quite a number of social phenomena which have some form of fundamental connection with activity (those of genesis, evolution, functioning), but at the same time, due to some of their features, they do not require address to activity and may be explained on their own terms without recourse to the concept. This fact leads to the problem of the completeness of the activity oriented explanation, i.e. how exhaustive it can be. The example of the cultural-historic concept in psychology shows that the explanation through reference to activity can be complete, or in other words, absolute. Here the concept of activity forms a unique explanatory basis for all psychological reality and this becomes a vital feature of the approach. In this respect L. Sev, as quoted by A. N. Leontyev, observed: in psychology "no conception based on the idea of 'a mover' inherently preceding the activity itself, is able to offer a basis and to provide a substantial validation of the scientific theory of the human personaltiy" [3, p. 66].

The situation appears totally different when attention is transferred to the other example I have chosen to consider, the study of science. On the one hand, an explanation with reference to activity happens to be very promising in this field and as I attempted to demonstrate earlier, it effectively reveals new strata of reality to study. But on the other hand, the other lines of the study of science unrelated to the concept of activity in explanatory principles (i.e. traditional historic-scientific analysis, various studies of the systems of knowledge) remain as important as ever. The situation is much the same in other fields of social cognition: linguistics, some branches of history, etc.

These latter circumstances must be kept in mind in the overall evaluation of the explanatory potential of the concept of "activity", since its universality sometimes leads to an overestimation of its potential to explain and to its application in situations requiring other means of explanation. This methodological extravagance has been fortunately avoided by linguistics, especially once de Saussure had introduced the idea of the conceptual distinction between language and speech, thus clearly establishing that the activity oriented explanation is but one among others. But in psychology, for instance, the situation is somewhat different. Sev's previously cited observation shows that activity can be treated not only as supreme but also as the only principle of psychological explanation. But this proposition, first, has to be thoroughly verified. The experience of the development of the cultural-historic school of psychology indicated that aided by the concept of activity one can efficiently explain the genesis, essence and fundamental trends of development of mentality and several phenomena crucial to psychology, such as motivation, human needs, personality. But does this exhaust the phenomena requiring an explanation in psychology? The question should probably be answered in the negative. As a matter of fact the premise that the human being becomes an individual only in activity and through activity by itself is not sufficient to infer that the concept of activity can explain all the manifestations of personality, the entire range of individual distinctions and so on. This conclusion is prompted by the simple fact that individuality is not only the product but also a prerequisite of activity, and so, at least in a certain sense, we can understand activity itself only through reference to the personality of someone who acts. Indeed, if we try to avoid it, it would mean the replacement of ordinary activity by Activity with a capital "A" where an acting being is a purely functional and consequently, in each particular instance, a purely optional appendage; which means that we are quitting the realm of psychology.

The insufficiency of the principle of activity as the unique and self-sufficient basis of psychological explanation is also corroborated by other facts. In engineering psychology, for example, the central problem is to explain not the development of the higher mental functions (this is instead one of the general prerequisites of research) but the real "performance" of already established properties and attributes of the mind in their interrelation [11], which apparently demands a somewhat different explanatory basis. Attempts to reduce the entire family of possible explanations to a single explanation based on the concept of activity encounter another substantial difficulty: how can the specific and real achievements of certain psychological schools and lines of thought be integrated with the activity oriented explanatory scheme? This problem still has to be comprehensively resolved.

These and certain other problems seem to be related initially to the fact that the constructive explanatory potential of the principle of activity in psychology has not yet been fully understood. This conjecture is apparently well grounded, and hence, the further development of the psychological interpretation of the concept of activity appears highly promising. However, one also has to keep in mind that there have been many cases in the history of science when an explanation based on one specific concept and claiming to offer an absolute and all-embracing solution has in the final analysis proved to be incomplete and relative. In any case it appears doubtful that the explanatory scheme of a specific scientific discipline can be reduced to a single concept, however prolific and structurally evolved it might be.

But this does not cast any doubt on the concept of activity in itself and its potential of explanation. We are only trying to be more explicit in evaluating this potential. To do this, it has to be realized that in science there can be no "ultimate" or final explanation covering all conceivable questions relevant to the given issue. Any concept, however universal it might be, poses quite definite limits to the subject of thought, and within this framework only some of the scientific problems may be resolved. This means that any research should acknowledge and make allowance for the limited scope of any explanatory principle or concept forming its basis. Or, to put it differently, there must be a limit to the extensive development of the explanatory concept, and when this is reached its further constructive application is possible only via intensive development, i.e. not by adding to the concept new phenomena and strata of activity, but by further elaborating the concept's inner structure, its meaningful content. But this brings us into an entirely new area, where the concept of activity ceases to play the role of an explanatory principle and itself becomes the object of study.

The objective content of the concept of activity should barely change with the modification of its functions. It is probably for this reason that the changes in its methodological meaning corresponding to these modifications of the concept function usually remain latent for a long period of time. Meanwhile in certain aspects these changes are extremely important. The essential point here is that activity as the ultimate abstraction designed to play the role of an explanatory principle does not need further explanations simply because of its very ultimateness (in each specific case this explanatory principle must only be methodologically validated as an adequate and sufficient foundation of explanation), whereas the study of activity as a special subject, on the contrary, requires a special explanatory procedure. If we, for example, use the concept of activity to explain the origin and formation of the mind, the concept itself needs no explanation - in problems of this type it is sufficient to present the concept in structurally developed form so that it supplies the meaningful mechanism for the solution of the problem. But if activity is studied as a special psychological reality, the integrity and conformance of that reality should be provided with a special explanation and validation, which obviously cannot be obtained via the concept of activity: otherwise, we have simply a tautology.

So in the later instance, the concept of activity is deprived of its marginal nature. Moreover, it appears that in this instance the concept itself needs a certain ultimate abstraction to explicate it. This problem requires special consideration; meanwhile, we will attempt to isolate the methodological conditions necessary for the concept of activity to be regarded as a special subject of investigation. The foregoing comments suggest that we are considering here two quite different ideal objects designated by the same term. The functional distinction between these ideal objects seems to be somewhat evident, and we would like now to try and trace the meaningful methodological consequences of that distinction.

The concept of activity in both its functions initially provides us with a reference to the reality being considered by establishing it within definite boundaries, and shaping it within a certain structure. In the first case, however, the boundaries of the reality in fact coincide with those of the universe of discourse which we are trying to cover via the concept of activity. But in the second case, the boundaries must be specially defined, to express it loosely: one can say that activity as the subject of study should be isolated from all else that is not pertinent to the subject and does not belong to the problems dealt with within it. The essence of this distinction is that activity as an explanatory concept can describe all the reality (viewed through this concept), whereas any specific subject is always a special projection of that reality: hence an explanatory principle may embrace the possibility of several subjects. This can be seen easily at both levels, that of the general concept of activity with respect to which concepts of activity in sociology, psychology, linguistics and other sciences are subject oriented projections, and the level of the activity concept interpreted in terms of a specific scientific discipline (e.g. different subject oriented constructions may correspond to a general psychological concept of activity as the explanatory principle – personality, consciousness, higher mental functions, etc.).

Another essential distinction can also be perceived in the structuring of the concept depending on its function. When used for an explanatory purpose, the structure of activity provides, as underlined earlier, the principal guidelines of explanation. In contrast, the structure of activity in subject oriented studies requires special validation; it cannot be regarded as a postulate or simply a convenient way of translating the general concept into the framework of research methods and techniques developed in the given field. This distinction is one of the challenges of the methodological analysis. Since the function of the concept of activity in each specific case is not explicit, the type and nature of the emergent activity structure are not discussed either. It might thus occur that a structure evolved within an explanatory framework may without any interim stages and consequently, without the required validation, be treated as the structure of the subject of studies. Repeated changes from one function to another may in the final analysis result in so much confusion that the real function of the given structural partitioning of activity is almost impossible to ascertain.

In psychology, a similar situation with the concept of activity has developed. The "double-series" scheme of "activity-action-operation" and "motive-purpose-condition" has been effective within the explanatory framework, serving the purpose of the psychological interpretation of the "object-oriented activity" concept to explain the genesis and meaning of the mentality. The aim of the structuring was to present all phenomena known in psychology as the product and elements of object-oriented activity. This is why action and purpose – the two categories already operationalized (particularly the former one) and with clear-cut psychological meaning – are respectively at the centre of both sequences. Action seems to express in concentrated form the psychological core of a far more general concept of object-oriented activity. In any case, the experimental psychological studies based on this scheme of explanation initially played around with the concept of the object-oriented action. One may even feel that the other two "levels" in this explanatory scheme (and we restrict our attention to this scheme alone), i.e. activity and operation, were used rather to further elucidate with the sole objective of facilitating the finding of an appropriate place for the subject of psychology among other sciences which study human beings: the concept of activity related the subject to the socio-philosophical disciplines and at the same time provided that wider integrity which produced and explained the object-oriented action; whereas the concept of operation closely linked psychology with neurophysiological studies, and because of this in certain cases even lost its purely psychological character.

In this way, the concept of the object-oriented action emerged as a starting point for the psychological interpretation of activity. But it is also the essence of a peculiar methodological ambiguity: it has become so intrinsically entrenched in the practice of experimental psychological studies that its very presence in the explanatory scheme has automatically made it a framework for a general psychological theory of activity, i.e. essentially the scheme of the subject of studies. Thus, one scheme combined two different methodological functions, and has to all intents and purposes been developed as a bifunctional one ever since.

In general, this fact by itself does not contain anything requiring a positive or negative evaluation. The heart of the matter is that regardless of its methodological bifunctionality, the scheme has always been and is still effective in psychology and has provided a solid foundation for obtaining new scientific results. This all attests to the indisputable scientific validity of the activity scheme in psychology. Doubts as to this scheme arise along quite different lines, involving its incompleteness or insufficiency for certain psychological research. In this case, the researcher is compelled to inquire what specifically is the source of this impairment of former efficiency - the scheme of explanation or the subject scheme.

True enough, the problem is stated somewhat differently in research. Because of the methodological disparity of the two schemetypes considered, one strives first to elaborate further the structural presentation of activity as a subject of investigation. This improvement may proceed along various dimensions. In [11], for instance, the persistence of the concept of object oriented activity as the explanatory concept is specially emphasized, whereas the scheme of activity as the subject of study is transformed into the four-level scheme because of the introduction of the level of functional units. It should be pointed out that this manner of transformation has been prompted by specific problems of engineering psychology: unlike the other lines of the psychological analysis of activity, in this instance operation is regarded in terms of its purely psychological origin; its role is no longer auxiliary but central and, consequently, a more detailed presentation that is lacking in the three-level scheme is required. The example also illustrates the relative independence of the concept of activity in its role of an explanatory principle and the activity as the subject of investigation.

Another line of development of the concept of activity in psychology was pursued by A. N. Leontyev [3]. Two points of his comprehensive analysis are of special interest here: first, his identification of bottlenecks (for instance, the issue of goal-setting) that have not yet been given a sufficiently complete and convincing explanation within the framework of the three-level scheme; and second, his indication of the necessity of a systems approach to the study of activity. But it should be noted that Leontyev thinks it unnecessary to review the structural scheme of activity, maintaining that it is probably sufficient for the further development of the psychological theory of activity.

Therefore in reality, the same structural representation of activity can perform at least two different functions at the same time. This creates, as we were able to see, certain methodological problems which are especially pertinent when the notion of the subject of investigation has to be improved and developed. Nevertheless, the duality of functions by itself should not be considered as anything abnormal. In a certain sense, it is even appropriate. Indeed, in the final analysis any scheme serves the purpose of explanation regardless of whether it is used as a structural frame of the subject of study presentation or as the explanatory principle. And if this functional distinction is expressed vocally, the objective is to reveal the difference in ways of life and development of the respective subject constructions. To all appearances the explanatory principle has an extremely wide field of application. Serving basically one main purpose, i.e. to relate certain mental constructions to a certain fundamental abstraction which determines the way of interpreting and structuring the reality, it turns out to be methodologically unassuming, the only requirement being that it supplies the means of understanding and organizing the reality within a certain framework or an entire set of subject-matter (but this does not imply the inherent simplicity and unobtrusiveness of the explanatory principle). This is why one explanatory principle can provide a basis of study for several different subject-matters, though not without some subtle modification in each specific instance.

As for the scheme of the subject of study, it retains its universal nature only within its specific framework and supplies valid answers only to questions relevant within the framework. This is the reason why the structural scheme of the subject-matter turns out to be flexible and changeable and, most important, does not exhaust the subject-matter construction as such. It should be noted that the concept of the subject of study has hardly been dealt with in methodological literature: both the firmly established ideas of the typology of the scientific subjects and the minimal set of the required attributes a subject should possess are conspicuously absent from our literature. But concerning our current field of interest, at least one property is seen to be obligatory for a scientific subject: it should provide the reality being studied with a certain dimensionality, not necessarily in terms of quantitative measurements (this is an attribute of only a certain type of subjects) but by singling out of that reality those formations which enable it to be presented as logically homogeneous, i.e. by defining a certain scale and frame of reference for it.

The design of the units of analysis representative of a given specific subject is a special but extremely important example of providing dimensionality of this nature.

The history of the study of activity as a special subject shows that constructions of this kind play an important and often decisive role. In linguistics, for example, once speech came to be considered as a special subject of verbal activity, attempts at analysis have been made to specify the intrinsic units of this activity that would embrace a sufficiently complete characterization of the act of communication as a linguistic phenomenon. Similarly, in sociology the concept of social action was introduced, which was used as the analysis unit in the study of a rather wide range of sociological phenomena. The same problem also arises in psychology, although there the situation is more complicated.

As was underlined earlier in psychological studies oriented towards the subject of activity, the concept of object-oriented action played (and to a large extent still does) a central role. In fact, it was this concept that was considered to be the principal unit of analysis. However, its status is definitely not clear, especially in respect to its claims to the role of the analysis unit. On the one hand, in psychology - and again one is not alluding to all psychology, but only to the line of studies concerned with the concept of activity - the concept of operation is also used as an analysis unit along with that of object-oriented action. On the other hand, in the three-level structure of activity, categories such as motive, purpose and condition occupy a special place; hence the conclusion that the motivating, "driving" components of the mental process are not accounted for by the units of analysis and, hence, should suggest a specific range of units. This leads to a particular situation where there is little choice but to introduce a psychological taxonomy to standardize a system of units required within the framework of this largely integrated line of psychological thinking. Otherwise, it would be impossible to speak of a unique dimensionality in the system of psychological studies. Besides, the concept of objective-oriented action as a fundamental unit of psychological analysis requires further elaboration and a more profound inner structuring: first, its permanent composition is not yet clear (which in turn is the result of the insufficient psychological development of the category of object orientedness); second, there must be a more precise and rigid concept of the stable relations of the object-oriented action with activity and operation; and third, a similar system of relations has to be established for the motivational components of intelligence.

I have touched on only a tiny number of problems accompanying the transformation of the concept of activity from an explanatory principle to the object of scientific study. The primary objective of this preliminary analysis has been to show that the transformation of activity into the object of investigation can be achieved spontaneously only to a certain point without systematic control of the methodological reflection. In the process of elaborating the new and quite unusual subject, which activity certainly is, there is no need to take any time to discover that the initial boundaries of the subject have to be revised, and that because of the wide range of activity manifestations at a certain stage, the theoretical analysis loses its logical homogeneity. All this has an impact on the integrity of the subject of investigation, stripping it of its prior wholeness and structural completeness; in this way additional grounds for the integrity are required. This is the main reason for introducing the principles of a systems approach in the study of activity.

While considering the introduction of the ideas of a systems approach to studies of activity, it has to be borne in mind that this combination is designed to elucidate activity as the object of investigation. For activity as the explanatory principle, systems ideas are hardly necessary; activity itself provides the logical centre of the speculation, although systems ideas may shed quite a bit of light on the explanatory functions of the concept of activity (showing, for example, that the universe of discourse determined by activity can be regarded as a system with respective meaningful characteristics of its relations and boundaries). But even if such is true, this does not present any exceptional difficulties for methodological analysis.

The application of systems ideas in the design of object-oriented constructs is of far greater interest in the context where the aim is to study activity itself as a special subject. But to what purpose? It is reasonable to assume from the outset that any subject of study which develops successfully does not require the introduction of new methodological concepts which would cause an essential transformation of the subject-matter. Hence, a systems

approach is advisable when the current subject of study cannot furnish a
satisfactory answer to at least certain fundamental questions. In other
words, a systems approach is usually used when a substantial methodological
restructuring of the subject of investigation as currently conceived is
required. But what is the direction of this kind of reconstruction and what
potential does the systems approach furnish?

It is common research practice to justify the utilization of a systems
approach by the implicit conviction that its ideas can help one to grasp
the elusive integrity of the subject and the reality behind it, to overcome
the overcomplicated pattern of the subject, and integrate its components,
however diverse they might be. This conviction - though often naive - is
not altogether groundless, since there are examples of application of a
systems approach application indicating that it definitely can do this.
But behind these examples lay the not always obvious research efforts, which
are indispensable for any methodology to produce positive results.

It should be emphasized that the systems approach, like activity, should not
be regarded exactly as a certain kind of subject-matter. It only refers to
a certain subject or rather to the reality stimulating the constructing on
that reality the object of a certain type. In methodological literature, it
has been oft-emphasized that a systems approach has nothing at all to do with
merely systems nominations. Analogously, the activity approach is not in the
least an apparently profound indication of human activity. And it is
examples of the constructive use of the "systems" concept which indicate that
its implication goes far beyond the meaning of the word "system".

One may recollect here that after Mendeleev had conceived of the idea of the
periodic system of elements, it took some time to create the systems object
before a method satisfying both logical and ontological criteria was found
to design a system. Similarly, a great deal of time elapsed from the time
the Gestalt properties were suggested by C. Erenfels to the first conception
enabling the interpretation of that phenomenon in terms of the systems
approach. These examples are quite indicative, as they illustrate a profound
and complex speculation linking the idea of a systems approach and the
systemic object. The examples reveal the methodological nature of the
systems approach. In fact, if this approach by itself does not identify an
object of investigation directly, it plays the obvious role of an
explanatory principle - this conclusion can serve as a starting point in the
process of elucidating the relationship between the systems approach and
activity.

This relationship is feasible in two forms: (1) activity and systems approach
coexist as two separate explanatory principles; and (2), the systems approach
functions as the explanatory principle with respect to the activity as
subject of study. The first type of situation is extremely interesting in
terms of the more general problem, i.e. the possible relationship of two or
more explanatory principles covering the same material (specifically, one
can assume that the meaningful priority in the pair of "activity-systems
approach" is a component of the activity principle, whereas the systems
approach is used as an instrument for the additional decoding of the content
of the principle of activity and the guideline for the particular arrangement
of the activity universe). This problem, however, should be examined in its
own right, which goes beyond the topic discussed. Here we will touch only
briefly on the second type of situation, i.e. that where activity and a
systems approach turn out to be functionally heterogeneous.

The range of application of using the systems approach to study activity as a special subject is determined by the fact that in its evolution, there has not been any particular necessity for the concepts and principles of the systems approach. The systems approach to activity is usually suggested only when the need to deepen the current understanding of the subject has been clearly realized. Unfortunately, there is often a rather simplified idea of these situations. In psychology, for example, appeals to apply the systems approach are sometimes supported by the implicit assumption that the systems approach would provide a desirable superstructure over an already structurally developed subject, and that this superstructure would not interfere in the least with the already achieved integrity of the subject-matter.

To accept this assumption would mean that both a systems approach and the intentioned subject to be constructed using something vague, uncertain and convenient rather than imperative (this might be the reason why calls for a systems approach are most often left until the final paragraphs of papers on activity).

The situation becomes even more complicated once the level of general declarations of the usefulness of the systems approach is exceeded. Since the subject of the system is not developed from the beginning the first thing to do is to critically reappraise the subject in its prior form and to determine in what respect it fails to meet the expectations arising from the changed scientific problems. When speaking about the necessity of a critical reappraisal we are alluding not to erasing the subject but rather of the imaginative thinking, oriented to analysing its most fundamental premises as well as its methodological relevance.

The constructive nature of this reappraisal depends critically on the initial choice of methodological commitment, which in our case is the systems approach. From the very beginning, the subject in its current form is to be embedded in the context of a new explanation. This, on the one hand, should record the specific nature of activity as a new reality and, on the other, should explain this reality via the concepts and mechanisms of the system formation (it is at this point, incidentally, where the difference in the explanatory status between the concept of activity and the systems approaches emerges: the former is of explicitly substantial nature, whereas the latter is rather more oriented to the laws of arrangement of form).

At the stage of critical reappraisal, the principles of the systems approach are used primarily to obtain negative definitions for the available subject: the systems approach has already defined new and more extensive wider boundaries for the scope of speculation, and has introduced a new dimension with a different frame of reference; hence the old subject is initially redefined within the new parameters and boundaries, which are used precisely for the purpose of finding out what it lacks. This is a highly relevant point, since it presents an absolutely indispensable prerequisite of a genuinely constructive use of the systems approach that is so often neglected. This negligence might not only undermine the continuity in the development of the scientific subject, but also distort the transparency of the specific directions of its restructuring.

One of the lines of reorientation of the psychological analysis of activity consistent with systems ideas might be presented in the following way. From the outset one should have a precise understanding of the objectives of systems explication of activity. Here two different situations are possible: this explication may take us beyond the realm of psychology towards a broader subject embracing sociological, economic and other aspects; or else, it keeps within the scope of psychology. A general attitude to this methodological dilemma is often somewhat biased because of the popular misconception that a systems approach must necessarily result in the design of a synthetic subject-matter. In fact, however, this is not quite the case [9]; anyway, we have every reason to believe that the fruitful application of a systems approach is quite feasible even out of the interdisciplinary field. This is also true of the psychological analysis of activity.

So if we decide to stay within the framework of psychology in our systemic explication of activity, we are first of all confronted with the fact that activity as a subject of investigation has not yet received comprehensive psychological definition. This is quite clear from the fact that certain critical aspects of psychological phenomena are sometimes compiled so that they can be explained by direct reference to economic or sociological concepts and notions of activity (e.g. division of labour, laws of commodity production). And it is not the methodological rigorism which is being examined here - under certain conditions economic categories also can or, probably, even should be used in psychological studies. But to serve the purpose of the psychological object, they must be psychologically reinterpreted. Failure to comply with this rule may result in the theoretical reconstruction of activity as an aggregate rather than as a system.

Within the same systemic perspective, it becomes clear that quite a number of fundamental psychological concepts still do not have a convincing interpretation within the framework of the psychological theory of activity. In this regard, the first thing that is conspicuous in a certain individual or consciousness here, for instance, are defined entirely in terms of activity. This is quite justified and even unavoidable once activity is used as a means of explanation, especially when the genesis and development of the personality or consciousness are being discussed. For activity as a subject of systems studies, however, this is inadequate.

The systemic interpretation of the concept of activity object-orientedness in psychology is exceptionally interesting. Object-orientedness should be conceived as the most important systems-forming factor of activity. At the same time, it is easy to see that in psychological studies this concept appears either as a philosophical-methodological postulate, or as the reality identical with the motive. It can be assumed that the constructive potential of this concept is far more extensive.

Finally, the systemic interpretation of activity poses a problem that has already been mentioned above - that is, the problem of the ordering (in fact of designing it anew) of the psychological taxonomy, of the identification of structural relations between different units of psychological investigation, and of checking their adequacy within the context.

Thus, we have briefly considered only a few elements of the systems approach to the analysis of activity in psychology - everything was necessary to characterize the methodological efforts involved. The above forms only an initial level of the application of the systems approach as an instrument for rethinking of the entire scope of studies. The stage of constructive critical reappraisal of the subject from the systemic standpoint should be followed by the positive determination and development of the systemic subject-matter. At that stage, the boundaries of the reality presented as a system and the principles of its systemic arrangement should be determined, or to put it differently, positive efforts should be made along the lines indicated by the reappraisal. This is the best way to fully utilize the explanatory potential of a systems approach.

In this analysis I attempted to show that a systems approach to activity investigations, as long as it is not treated simply as a gesture to methodological fashion, is a genuine challenge requiring that the scientist rethink both his methodological and substantive ideas.

I have concentrated primarily on the methodological aspects of the rethinking process, on what could be called the methodological framework of activity as an object of study. The most important conclusion of our analysis is probably that activity as a subject in general does not simply present itself to the researcher in a straightforward way, but demands to special reconstruction as a special ideal object, along with the determination of various functions of the concept of activity. This is especially true of a systems approach to studies of activity. In this particular case, any positive results can hardly be expected without any major preliminary methodological effort.

REFERENCES

1. G. Allport, *The Personality*. New York, Holt, Rinehart and Winston (1937).
2. V. V. Davydov, *Modes of Generalization in Learning*. Moscow, Pedagogika (1973), in Russian.
3. A. N. Leontyev, *Activity, Conscientiousness and Personality*. Englewood Cliffs, Prentice-Hall (1978).
4. M. K. Mamardashvili, *Forms and Content of Thinking*. Moscow, Vyshaya Schkola (1968), in Russian.
5. A. P. Ogurtsov and E. G. Yudin, Activity, in: *Great Soviet Encyclopedia*, Vol. 8. Moscow, Sovetskaya Encyclopedia (1970), in Russian.
6. M. S. Rogovin, Development of the structural multi-level approach in psychology, in: *Systems Studies. A Yearbook-1974*. Moscow, Nauka (1974), in Russian.
7. V. S. Shvyrev, The problems of elaboration of the activity category as a theoretical concept, in: *Metodologitcheskiye Problemy Issledovanya Deyatelnosti VNIITE Proc.* Issue No. 10. *Methodological Problems of Activity Studies*. Moscow (1976), in Russian.
8. L. S. Vygotsky, Development of the Higher Mental Functions. Moscow (1960), in Russian.
9. M. G Yaroshevsky, *Psychology of the XX Century*. Moscow, Polytizdat (1974).
10. E. G. Yudin, Methodological Essence of the Systems approach, in: *Systems Studies. A Yearbook-1973*. Moscow, Nauka (1973), in Russian.
11. V. P. Zinchenko and V. M. Gordon, Methodological problems of the activity studies in psychology, in: *Systems Studies. A Yearbook-1975*. Moscow, Nauka (1975).

BASIC CONCEPTS AND DEFINITIONS IN SYSTEMS THEORY AS APPLIED TO BIOLOGY

A. A. Malinovsky

Definitions of basic concepts used in systems research differ greatly. Many authors using a similar term often imply entirely dissimilar things [5]. One finds that even in cases when the concepts are defined and can be compared, they do not meet certain criteria. What are these criteria? The decisive one is, of course, the practice of application of a given concept. However, any concept must also be substantiated logically. It is this logic context within which one must seek to define at least basic concepts such as "system", "wholeness" and "organization" to ensure that these concepts are generally acceptable.

Since these concepts were made use of earlier to deal with specific research problems, one can speak of certain arguments indicating that these concepts are justified as concepts at least in biology and related sciences. In particular, they find application in problems such as the adaptive role of genetic and reflectory mechanisms, optimal couplings within a body, the evolution of endocrinal mechanisms, optimal population structures, alternative physiological processes, feedback in (swarm) regeneration, the role of various types of feedback in embryogenesis and differentiation, and feedback in various types of abnormal processes and analysis of their origin.

Ackoff [1] was one of those who attempted to establish basic concepts and definitions for systems research. In his paper, Ackoff tries not so much to elaborate general concepts (such as "system", "structure", "order", "organization" and "wholeness") or to determine their interrelations. Instead, he considers more special types of systems and attempts to give, in the long run, a general description of highly organized (adaptive or learning) systems. As a result, he overlooks many important system features. Right in his initial few definitions, Ackoff follows the generally accepted view of treating the interrelation of system components as the basic identifying feature of a system. Though not explicitly, he in fact implies that component interrelation is stronger than the relationship between system components and the environment. He does not give general definitions enabling systems to be classified, but rather a classification as such, i.e. the classification into closed or open; static, dynamic, or homeostatic; on-line or off-line, adaptive or learning systems.

Ackoff abandons the attempt to define several of the most basic and
elementary concepts necessary to develop system theory and instead engages
in a more detailed and relatively narrow systems classification. He
neglects the basic problems in systems theory, namely relationship between
the system structure, types of system development and system functions, and
describes only system functions irrelative of the system structure.

On the other hand, a great many system features which methodologists in
systems research generally fail to note were pinpointed in Ackoff's paper.
He writes, for example, that systems exist which reduce the diversity of
systems and which are less organized than at least some of their constituent
components. In contrast, many authors believe that any system is more
efficient and better organized than any one of its components or any simple
(not combined in a system) set thereof.

More general and more important is Ackoff's contention that any system has
an unlimited number of properties, with only few of them being of importance
for any given problem [1]. Of course, this statement by no means contradicts
the objective nature of system laws and systems research techniques. It is
similar to quantitative techniques of studying a given object as the
quantitative characteristics of objects (and object sets vary greatly from
one problem to another).

Finally, Ackoff's paper contains the important suggestion that it is no use
to improve any part of a system if this improvement leads to a deterioration
of another part to the extent that it outstrips the above improvement [1].
This statement is similar to the "weak link" principle proposed, by
J. Libich specifically, for the conditions of plant growth. However, Ackoff
does not classify systems by laws of motion, and as a result he overlooks
the fact that this statement only applies to one particular class of systems
(I will denote them as relatively "stiff" systems) while it is inapplicable
to "corpuscular" or "discrete" systems such as an animal species which is
essentially not weakened as a whole by weakening or eliminating (of course,
to a certain extent) some individuals, as other individuals only gain from
the reduced intraspecies competition and quickly replenish their number by
the multiplication process. However, Ackoff's latter statement refers to
specific laws of systems behaviour (in fact, as can be seen from the above,
only to a certain class of laws) rather than to general definitions in
systems theory.

Many authors apply the concept of "system" only to objects of natural origin
(in fact, to stable ones), i.e. to objects which are organized in some way
(see [7]). This limitation is not found necessary, however, as systems are
possible which are produced artificially or which are quite unstable and
exist for only a short span of time (between conception and quickly
approaching decay). Other authors consider the presence of strong ties
inside a system as its central identifying feature. This can be viewed as
the system wholeness requirement. Here again, no significance is attached to
whether the system was conceived naturally or artificially. In other words
this definition of the system includes the simplest and generally accepted
fact that a system should include a set of interrelated components, i.e. that
any system should have a structure, while it also implies some additional
requirements such as organization, wholeness and stability - which we find
unnecessary. An actual system may or may not have any of these properties,
as otherwise it would be unnecessary to single them out as separate items.

It appears more correct to separately treat any property that may or may not be present in a system. In considering any given system, one has then to employ the concept of system as neutral in terms of that specific property before the system can be more fully described in terms of the presence (and the degree) of order, wholeness, organization, stability, etc. Indeed, the many obviously highly organized yet unstable systems will be excluded from consideration once stability is implied to be a necessary condition. Thus, males of some Arthropodae die after they have fulfilled their fertilization function, that is, they are unstable. Yet it is clear that they have a rather high degree of organization. Equally, a disorganized system is generally possible, i.e. one in which some positive properties of components cancel each other out so that the net amount of these properties in the system is less than in a set of its separate components. Even wholeness is a highly conventional property from the standpoint of a general definition of system. For example, one can regard as a system a set of objects related by similarity (e.g. a species which can be found in different zoos throughout the world but is extinct in the natural state) or by an accidental time and space conjunction that will later be destroyed. Nevertheless, systems of this kind do exist and have certain properties.

Furthermore, the most general and basic concepts in system theory should not overlap and should be strictly delineated. If, for example, organization is introduced as a separate concept, it should not necessarily be included in the scope of the concept of "system". In fact, this inclusion is inevitable in some cases; however, it should be effected in a form which accounts for the hierarchy of concepts involved, so that one of them is narrower than the other and is part of the latter. Thus, organization in itself implies a certain order, i.e. the former is a special case of the latter. Therefore, the organization concept enables one to discriminate between ordered systems which have the organization feature and those which do not. However, the "system" concept is the most general, and therefore we think that it should not include concepts of order, wholeness or organization.

In addition to the requirement that any system contain at least two components, one must assume that any system has some structure. We will call *structure* any relationship between components and their interrelation within the system. Unless these relationships are determined in detail, they may have zero values; that is, they may have no meaning other than that they are just considered to be integrated into a system. The term "structure" thus denotes a mere description of any type of relation and interaction between components inside a system, to include the most complex types of wholeness and organization.

Now turn to the definition proper of some basic concepts.

There would be no point in providing a special definition of the concept of *component* at this time since it has been widely used in various sciences and is neither new nor disputable. This concept denotes those parts of the system which are not subject to decomposition in the subsequent study. Whenever the term "component" is to be used in a new meaning, this usage is specially introduced (e.g. in connection with the concept of subcomponent).

We will call a *system* any set of material, energy or information related components singled out of all the other components of the environment according to a certain principle. It is not necessary that the system behave differently from the sum of its components, each taken separately. In fact, a difference of this nature often exists, but then again it may not. A difference in behaviour appears provided that the components are somehow organized within the system. If the level of organization is low, the behaviour of the system is similar to that of a simple sum of its components, while where there is a high level of organization, the system behaviour is quite different. Of course, the more intimate interrelations between groups of components, the more reasonable it is to combine these groups into a natural system. However, whether a system is organized or not depends more on the nature of these ties specific to any one type of organization than on the degree of strength of the ties.

In a paper by A. I. Uëmov and myself [3], a definition is given which uses the degree of the strength of component ties as a criterion for identifying a system. While applicable to most systems of natural origin, this criterion is not obligatory for every system, particularly artificially identified systems.

In studying only theoretical aspects of a naturally developed object or phenomenon, we can identify systems according to the "natural ground" delineating them from the wholeness principle. This principle states that any coupling between components inside a system is stronger than any coupling between a component of the given system and a component of some other system. This opens the way to the independent treatment of each of these systems. However, a different delineation of systems and/or components is equally possible.

Here a certain similarity with quantitative research can be traced. The number of relatively independent organisms is initially calculated when studying certain objects, such as animal populations. However, under certain circumstances, a separate count may only be required of some fractional components present in each organism of a population, for example, in order to study the energy consumption or energy release by the entire population. The problem might be the temperature increase in a beehive compared to the environment, though this difference is determined by the activity of all the bees belonging to the given hive. One then considers the energy systems of the entire population, and does not delineate natural boundaries between individual organisms. One may also consider a certain material indicator, e.g. the production of honey in the hive, and evaluate the amount of honey while neglecting questions such as how many species live under these conditions and provide for the accumulation of that honey or how much honey and for what period of time each bee yields.

A similar situation arises in employing the systems-theoretical approach to solve a particular problem. If the aim is an approximate, unbiased and "neutral" study of a population, the latter can be treated as a system of individual organisms each representing naturally developed and relatively autonomous components of the population. On the other hand, boundaries are drawn between subsystems or even subcomponents of the population if special groups of these organisms are to be studied. Thus, the family can be identified as a population component. However, only certain functions of the organism are taken into account, while its morphological parts are neglected, if types of behaviour of the entire population are to be studied. These functions can be analysed irrespective of whether only one (and what particular) organism or many organisms display a response (such as an alarm signal) that is important for the entire population. In other words, the "research cross-section" of the population will in this case be quite different.

It follows that one must differentiate between (1) the simplest and approximate ("neutral", to put it conventionally) approach to a system treated whenever possible in terms of its wholeness and independence of the environment and (2) specific versions of systems-theoretical approaches to various objects depending on the particular problems posed and with the objective of solving a specific problem. In the latter instance, the system is resolved into components depending on the problem statement, and not at the points where the ties are weakest; that is on a quite different basis matching the problem involved.

Structure is an inherent property of a system. It is the link of a definite type between system components. A structure can be stable or unstable, while a tie can be static or dynamic, direct or indirect, unilateral or bilateral, etc. The types of ties unifying a system can be described as a special type of structure. The structure is often identified with the system. However, I believe that these concepts should be clearly distinguished. A structure is an abstract type of tie independent of the number and quality of the components. It can be conceived as a characteristic feature of a cross-section of any given system at a given time. In contrast, a system is a specific set of components, with a definite type of functioning and development for any morphological structure.

I believe that the primary goal of systems theory is to establish (1) which properties of a system relate to its structure and which to the number and quality of components and (2) what relation exists between the type of structure and structure properties (i.e. the functions and laws of motion of systems with a given structure). One must discriminate between the internal structure and the external structure [7]. Inside any system, such as the design of any machine, there is a certain type of tie between the heat source, the vapour pressure, the piston motion, the transmissions, the control systems and so on. In contrast, the external structure is represented by ties between external features of the system in terms of their relationship both to the environment and between themselves. Thus, when the behaviour of a mechanism is being studied, external features will be related to the environment and will determine the ways of coupling the system to the environment, the system response to the environment, and the system impact on the environment. Internal ties are represented by certain quantitative and qualitative laws of action of some part of the system on another depending on the role of these parts, the type of transmission, etc., while the external ties of the system are determined by their interrelation and by their relation to the environment.

The concept of *order* requires no definition, as it can be reduced to the well-known structural negentropy: a deviation of component distribution from the most likely one under given conditions. This means that order is a quantitative parameter of the system structure. Order can be the same for different structures, since it is only used to evaluate a single feature of the structure in question, namely, the deviation from the most likely ratio of components which are random under given conditions. On the contrary, similar structures always have a similar order. A certain order is also the *sine qua non* of wholeness and organization. Yet order itself is neutral, in that the presence of order does not imply wholeness or organization, these being characterized by the type of order, i.e. by the structure rather than by the degree of order. An increase in structure negentropy as such does not imply either higher or lower system organization. Order (the structural negentropy) may take on various values and yet, a system having the same order may or may not be whole, and may be more or less organized or disorganized, since the nature of wholeness and organization is determined by the specific type rather than by the degree of order.

Organization will be defined as a measure of distinction between the properties and the behaviour of the system on the one hand, and those of a simple sum of the system components on the other. We can assign either a positive or negative value to this difference depending on the particular problem. The organism of an animal or a plant is, for example, a highly organized system in terms of viability. Its components could not exist independently (this is true at least of higher forms). In a certain combination and with a certain degree of organization they produce, even in lower forms, an essentially higher viability effect than the sum of these components taken separately. The same applies, for example, to a symbiosis of organisms. Two organisms combined symbiotically drastically increase their viability in certain cases. Stability, however, is not always a positive characteristic.

Organization is thus a relative measure. It is often related to the viability and stability of an object only at an early stage of study. This criterion may be replaced by another when a solution to a specific problem is sought. In fact, even in terms of viability, any given system is better organized in one respect that in another. The organism of a lower species is highly organized in terms of physiological information. It is better adapted to the environment than the human organism, while the latter is better organized as a result of the evolution of the higher nervous system and because of the ability to unite into social communities. Order is an abstract measure representing the system as a whole, "impersonally", as a mere number while organization is a specified, "denominated" measure which for any given system can vary according to the standpoint.

As well as possessing different values, order can have different degrees. This means that organized systems acquire new properties in a greater or lesser degree. Essentially, there is no system whose components lose all their characteristic features once they are integrated. It would therefore be imprecise to say that any organized system differs in its parameters from the sum of its components [6]. While acquiring entirely new qualities not present in any one of its components, any system inevitably retains to a certain extent the properties of these components not only in a transformed form but, in respect to certain parameters, in the most direct sense.

Wholeness is, in my opinion, a special case of order and organization. A system will be called whole if all its parts function (and evolve) together while the entire system is independent of the environment and other similar systems. This independence is, of course, only relative, and implies that the system must somehow respond to the influence of other systems. This response does not amount to component parallel evolution (as is the case with the growth of body organs or species evolution) and is rather a reply to an external stimulus. This situation develops, for example, when two confronting species (i.e. a predator and its victim or two competing species) interact directly. Wholeness is therefore not determined by the level of organization, but rather by the degree of tie "density" within the system, this "density" being higher than the "density" of ties between the system and the environment [3].

Relationship between the structure and its components. A comparison of different systems composed of different components but with a common structure permits the identification of their common features. All discrete systems of homogeneous components of the corpuscular aggregate type have certain features in common. Similarly, systems organized on a centralization principle (whereby most entities are united because of their relationship to a central entity) have common features, even though there are also some distinctions.

Thus, whenever a common character is present in the structure of systems, it produces a specific pattern unifying all these systems even though their components may be quite unlike each other. It would appear then that structure has an all-prevailing significance independently of the specific properties or number of components. Though the number and qualitative features of components which are able to form a particular system can vary over an extremely wide range, this does not mean that they have unlimited applicability. In other words, a particular type of structure can originate and evolve to integrate components that are quite unlike each other in terms of both quality and quantity, yet certain qualitative and quantitative limitations are always present.

Structure levels can be delineated on the basis of different principles. Levels can in certain instances be identified, for example, according to the degree of system complexity. Any lower-level system is then a building block for a higher-level system. Thus, molecular components make up cellular formations, these are components of organs, organs constitute an organism, the latter emerges as species in a part of the biocenosis on the one hand and of a larger systemic entity, on the other (from the standpoint of systematics), and finally all the living formations on the globe are parts of the entire biosphere. This upward movement across hierarchical levels corresponds to increasing levels of organization. Cells of a body are in a sense better organized than certain multicellular organisms considered as entities. On the other hand, the entire organism can be said to have a higher organization than its constituent cells, in that the organism as a whole has several properties absolutely impossible at the cellular level. At a higher level a certain loss of organization appears. The species composed of individual organisms has certain advantages (e.g. plasticity and a longer life span) over the organism, but are simpler as far as structure is concerned. Viewed as a system of individual organisms, the species is, of course, simpler than the organism as a system of cells and organs. In fact, the species as a whole loses some of its advantages over individual organisms. Highly organized animals as a whole lack the intellectual abilities that an individual organism has. While some highly organized individual animals (such as antropoids, the elephant and dolphin) can pose complex goals and achieve them through a pre-planned action, this is beyond the capabilities of the species as a whole, the latter developing at random and through natural selection. J. B. S. Haldane said that selection is shortsighted as it does not involve achieving a preset goal such as the sacrifice of a victim at the next stage of action. It follows that a higher hierarchial level does not necessarily imply a better system of organization.

The treatment of organization levels can follow either of two directions. On one hand, certain criteria can be introduced to describe system components and the organization of the system as a whole. Systems may be classified as passive, active, self-controlled and self-generating. (Apart from self-control, a goal-oriented evolution is possible in the latter.) On the other one can use an empirical approach to this problem and treat systems merely in terms of the effectiveness of their organization. Only the degree of organization, that is, the effectiveness of a system as compared to that of the sum of its constituents, is assessed from the knowledge of certain parameters. Here levels of organization effectiveness come into question.

The time evolution of a system can follow different patterns. The system structure may radically change and the components may be replaced during evolution. The overall composition of system in terms of both its components and structure at an early stage has little, if anything, in common with that at the final stage. And yet it is one and the same system. The organism of an animal or a plant consists initially of a single fertilized cell, but it subsequently evolves into a multicellular body with billions of cells. Its structures have differentiated and totally changed and even their constituents have become different. Yet a certain conformity does exist, implying that the entire evolution of a system is continuous so that any new stage is linked with the foregoing one, the former being, in fact, the latter transformed. It should be noted that time intervals can be made as small as possible to reduce the differences between structural stages of evolution.

In certain instances, a situation arises in which an original system disintegrates into two or possibly three new systems. As an example, when studying the divergence of species originating from a single species, the question is posed: with which of the daughter species should the original be identified? The answer clearly depends on the form in which the question is posed. If the final result, i.e. the daughter species, is considered, each of these species during its evolution over time can be identified with the mother species and treated as an integral evolving system. However, if we investigate the evolution of the mother species, we have to conclude that it has been destroyed by a disorganizing process. We can treat the daughter species depending on the statement of the problem either as evolution (if preservation of a major part of the genofund is of concern) or as a result of destruction (if the objective is to know how much of the species morphology and of the ecological relationships have persisted) of the original species.

I do not claim the above definitions to be complete, yet they might be used in subsequent research as basic concepts – not only in their own right, but in their interrelations. The above attempt to provide definitions for systems theory concepts will hopefully open the way to a more precise logical characterization.

REFERENCES

1. R. Ackoff, Toward a system of systems concepts, *Izv. Akad. Nauk SSSR: Tekhnicheskaya Kibernetika*, **3**, 68-75 (1971), in Russian.
2. I. V. Blauberg and E. G. Yudin, *Systems-theoretical approach: Development and essence*. Moscow, Nauka (1973), in Russian.
3. A. A. Malinovsky and A. I. Uemov, System types and principal biological laws, in: *Organism as a System*. Kiev, Naukova Dumka (1966), in Ukrainian.
4. A. A. Malinovsky, Science of organisation and organisation of science, *Priroda* No. 3 (1972), in Russian.
5. V. N. Sadovsky, *Foundations of General Systems Theory*. Moscow, Nauka (1974), in Russian.
6. Systems-theoretical approach to biology: materials presented at a panel discussion, in: *Systems Research Yearbook 1970*. Moscow, Nauka (1970), in Russian.
7. G. P. Shchedrovitskii, Classification principles for the most abstract trends in methods of systems research, in: *Problems in Studies of Systems and Structures: Materials for a Conference*, pp. 15-23. Moscow, Akademiya Nauk SSSR (1965), in Russian.

PRINCIPLES OF SYSTEMS RESEARCH AND APPLIED BIOLOGY

K. M. Khailov

PRESENT-DAY STATUS OF SYSTEMS APPROACH IN BIOLOGY

The last 15 years have witnessed a consistent and fairly rapid development of the systems approach to problems of biological organization and related research methodology [5,7,8,14] unprecedented in Soviet biology since Bogdanov published his book in 1925 [3]. The first seminar on the subject was held in 1968 and the number of relevant publications has since grown rapidly.

At first, some biologists felt that there was nothing essentially new in the systems approach as far as biological theory was concerned, nothing commensurate with the evolutionary theory. Others, on the contrary, pinned great hopes on it not only in the realm of theory, but also in the solution of specific applied problems. At present there are sufficiently tangible results of using systems ideas in most diverse areas of theoretical and applied biology. These results were subjected to a special philosophical analysis in terms of their methodological implications [2, 4, 10, 11, 16]. To a lesser degree, they were discussed among biologists themselves. A paucity of such analysis is all the more evident since quite a few biological (and particularly ecological) theoretical and applied problems are oriented toward multicomponent and multiconnected systems and systems methodology is widely used in their study. Meanwhile, some biologists are still sceptical as to whether the concept of a "systems approach" is sufficiently lucid and unambiguous. This is manifested, among other things, in the useful polemics, characterizing contemporary theoretical-biological discussion of systems ideas and principles.

It is certainly true that the theoretical framework of a systems approach is not always abundantly clear, flawless or effective. Yet, in my opinion, its application to biology has a number of fundamental advantages. Let us discuss these advantages, it being understood that the list could be different depending on the researcher's individual orientation and the specific area of biology.

(1) Systems ideas and methods have become an organic part of practically all branches of the life sciences - systematics, functional morphology of animals and plants, phyto-geography and zoo-geography, physiology and biochemistry and, above all, ecology - the study and control of a rational

use of natural environment. They figure prominently in tackling the most urgent global and regional ecological problems.

(2) Although the ideas of a systems organization of life proved attractive and constructive as such, the constructive character of the entire systems approach was largely determined by the development of a number of specific versions of system methodology that found wide practical application. Suffice it to mention mathematical modelling, methods of multivariate statistics in analysing complex systems, methods of mathematical design of experiments. Each of these varieties of systems study may give rise to some criticisms (there are even some doubts as to whether all these methods are "systems"), but their wide use and obvious effectiveness in solving problems, involved in a study of complex objects, speak for themselves.

(3) Practically everywhere where systems concepts and methods are applied, they have become organizing conceptual centres or have merged (as in ecology) with the conceptual framework of biology itself. They suggest to specific biological disciplines the need for interdisciplinary integration, giving top priority to the problem of integrity of a living organization and its functioning.

Systems ideas also perform a useful correcting function. For instance, a trend has recently emerged toward multiple biological studies, frequently associated with getting the maximum amount of data with the utmost precision. From the standpoint of systems studies, envisaging the collection of strictly defined goal-oriented data, this strategy is ineffective. In this sense, the systems approach is an organized and, therefore the best, version of a "multiple approach".

(4) What is important, however, is not only the organizing conceptual power of the systems approach. A specific systems methodology (as, for instance, mathematical modelling) is oriented toward pre-planned data collection and automation of measurements, data storage and processing. This stimulates the development and application of special technical systems to perform supportive and organizing functions in tackling complex ecological research problems [9]. An interdisciplinary approach to the formulation and handling of major biological, ecological, developmental and ecological-economic problems, attended by the development and use of special information retrieval systems leads to the emergence of what may be termed "a systems industry" in science, an industry where biology occupies an indispensable and important place.

Nevertheless, in reply to the question as to whether the forecasts concerning the potentialities of the systems approach to biology, made when it was first widely discussed, have fully come true, it could be stated that the most optimistic (or rather immoderate) of these forecasts have failed to materialize. Among these was the belief that an all-embracing "general systems theory" could be evolved, including a "theory of biological systems" which could form the basis of modern biology. What did materialize was the emergence, alongside the evolutionary conception of an equally – and quantitatively and theoretically even more thoroughly – developed conception of biological organization.*

---

*Neither the former nor the latter in my opinion, as yet qualifies as a theory.

This had a highly symptomatic impact on biological education curricula at our universities: the proportion of evolutionary conceptions has relatively decreased while that of the conceptions associated (in various disciplines) with biological organization showed a marked increase both in absolute and relative terms. Yet special courses in the organization and functioning of living systems are still few and far between.

At the same time, the systems approach cannot claim a leading place in biological research which is very good since any monopoly in science is bad. In my view, however, the present status of this approach, has not measured up to initial expectations. This may be partly due to the difficulties involved in mastering its empirical methodology, based on a wide use of mathematical methods and computers. But, on the other hand, this is an indication of its still insufficient methodological effectiveness. At any rate, it is clear that its possibilities, just as those of any other approach, are limited.

Having become, nevertheless, a successful tool, the systems approach has naturally given rise to complex and specific problems in biology. Therefore, in discussing its present status, it is more appropriate to describe the "working" difficulties, associated with it, rather than its efficiency.

We shall cite below examples of biological problems and difficulties arising from the use (or non-use) of systems principles.

SELECTION OF A PART OF A COMPLEX OBJECT FOR STUDY AND THE VALIDITY OF RESULTS

The methodology of studying complex objects has now been so thoroughly developed that an outline of just some general systems principles would sound naive to an expert. One thing is clear, however - and it is of little concern to systems experts though it matters a great deal to biologists: not only are most of the biological studies still "pre-systems" but, more often than not, the vast amount of material, collected by biologists and pertaining to complex objects, cannot be processed at all in a satisfactory manner. The effort, exerted in collecting it, does not pay in such cases. To some - perhaps, considerable - extent this is due to failure to comply with elementary operational rules, governing the acquisition of material. Whether we like it or not, these rules are part of systems methodology. One of the typical reasons for the difficulties, involved in processing any array of biological data, is that in collecting it the researchers select measurement parameters arbitrarily or to meet narrow professional interests.

If living nature were a loosely ordered set of organisms and species, no strict requirements could be imposed on the selection of the part for study and any principles of linking this part with the entire set would be out of the question. But what is studied in its multilevel hierarchic organization in each particular case is but a small part. The isolation of this part runs counter to the principal condition of the systems approach - the observance of integrity and no dismemberment of the object into isolated elements. Therefore, the best thing to do would be to study a whole organism, a whole community, a whole ecosystem, biosphere. Yet, the integrity principle does not apply as far as they are concerned for many objective reasons. The first reason is the complexity of these objects. The second is the "narrow" professionalism of the researchers' own orientation. The third and most important reason is the non-evident boundaries of many natural systems.

It is no coincidence that the definition of a species and its boundaries has figured so prominently in classical biology. Defining the boundaries of most of the ecological objects is still more complicated. The trouble is that their composition is based on a combination of, at least, two different principles - probabilistic, typical of diffuse systems with a relatively low functional diversity, and deterministic, typical of rigid systems with a high intrafunctional diversity. As a result of this dualism some of the elements of living (particularly, bioconservative) objects are permanently or periodically interrelated through a network of one-to-one relationships while other elements enter into optional relationships which renders the structure of such systems essentially ambiguous and makes it difficult to properly isolate units for their independent study.

The inevitable isolation of parts of complex objects has an important consequence: the functional characteristics of parts outside the system are essentially different from those within an integral system. This necessitates the linkage of the part being studied with the whole. In input-output models this linkage is of necessity always made but in other types of studies it is often omitted. Meanwhile, one of the major goals of this linkage is to determine the domain of valid results in accordance with the position of the isolated part. Needless to say, this is only partly true of diffuse systems.

The above propositions are trivial. Yet they have to be stressed because in numerous biological studies the object of research is not defined or selected at random in a way, not optimal in terms of research procedure. In such cases it occasionally happens that the results obtained are adequate in a very narrow field and cannot have the importance (particularly in terms of their predictive power) that the researcher frequently attributes to them. Not only the researchers themselves but also the authors of surveys, especially those given to broad generalizations, have to take into account the way each study complies with the operational rules of systems analysis. This, however, is not always done either.

When the structure and boundaries of a complex object can be established in a sufficiently unambiguous manner, the isolation of subunits for independent studies should follow certain rules, such as the following:

(1) The isolated group of elements should correspond as closely as possible to the structural subunits of a natural object, i.e. should not be a random combination. Therefore, the principal study should be preceded by preliminary qualitative modelling for a well-founded restriction of the problem and its simplification. Biologists do not comply with rule very often. A preliminary study may also be required for a rough estimate of the inputs and outputs of the blocks to be examined in a simplified manner.

(2) It is desirable that the breakdown of a system into interacting subunits should not be carried to the extreme - to paired combinations of elements (which is very often the case in empirical biological studies). At the pair level the objective choice of pairs that merit a careful scrutiny becomes a very complex and unprofitable task with the domain of valid results reduced to a minimum. A study of paired combinations within a complex system amounts, in fact, to reducing a large number of states to certain particular ones, selected from the system at random and having a very low cognitive value.

(3) In studying any simplified versions of a system it is necessary to strictly control the variables constituting immanent parts of a given natural object but not included in the part to be studied. Besides, fixed values of the variables must be included in the description of the experiment for

unless this condition is met the domain of valid results is indeterminate, and the results lose their scientific value. Nevertheless, biologists, fully aware of the essence of this principle, do not have much chance to comply with it for the simple reason mentioned earlier: the system boundaries are not defined and it is unknown which variables out of innumerable parameters should be controlled in the first place. It is obvious that if a system is broken down into individual pairs of features, the control of variables becomes an impossible task for the smaller the part isolated, the more other variables determine its behaviour and should be controlled in the experiment.

In many contemporary studies the utmost care is taken to observe the above operational rules. I will cite some typical examples.

A. The part being studied is limited but selected in accordance with the object's immanent structure. If, for instance, not the entire ecosystem but its biological block is selected with its main trophic levels and their subdivisions, the physical and chemical blocks (abiotic environment conditions) may be treated in a simplified manner. As each block is reduced, the validity of the model will be restricted but it will still have a considerable prognostic value. A sizeable part of present-day ecosystem studies are based on this pattern.

B. Suppose the problem involves a very important part of a natural ecosystem but a number of variables, affecting the parameter being measured, have been subjected to rule-out selection and the principal variables thus revealed have been singled out for further study. The remaining relevant variables have been fixed, their values determined and reported in the description of the experiment results, the latter being valid only at such fixed levels of the variables. The results are comparable to other data obtained at similar fixed variable levels. This case may be considered optimal for studies with few factors involved.

More often than not the operational rules of systems studies are to some extent departed from. In many instances this is due to objective reasons to which it is impossible or difficult to take exception. But sometimes this is caused by subjective motives compounded by misconceptions concerning the application of results. In order to understand how operational rules are departed from and precisely what consequences it may lead to let us examine the following hypothetical case.

Let us assume the researcher is an ecologist, well aware of the functional complexity and structure of both an ecosystem and an organism, but is guided by his own professional interests. Supposing he wishes to study the connection between ecosystem temperature and animal respiration or the connection between light intensity and photosynthesis. In both cases one element of the pair is related to the first organization level and the other to the second. Sometimes the choice of pairs is determined not even by professional interests but by the availability or unavailability of a certain measuring instrument. For instance, the availability of a salinometer and the unavailability of an oximeter may result in a study of the effect of environmental salinity on the organism rather than the effect of oxygen concentration, although from the standpoint of ecology the second factor may be much more important than the first.

But let us assume that precisely these pairs were selected. On formal grounds each pair may be considered a system. Both pairs, however, are equally remote from both an ecosystem where temperature and light intensity depend on many ecological variables and an organism where respiration and

photosynthesis are related to the organism's other vital functions and
parameters. It is therefore more fitting to say that with the problem so
formulated the study has no object of its own. In other words, the part to
be studied has not been selected in an optimal fashion. Non-optimality is
by no means an irrelevant feature: the results of studying such pairs,
selected at random, are valid only within narrow limits which makes them
uninteresting both to ecologists and morphobiologists. The concept of
"control" as used in studies involving few factors loses its original sense
in a multifactor system. Data processing by the methods of univariate
statistics, a panacea according to a widely held belief, likewise makes less
sense.

The selection for study of a certain group of elements in a complex object
is frequently arbitrary, too. The taking of measurements with the utmost
accuracy is intuitively felt to be correct. However, if the pair is selected
for study at random and in isolation from other relevant variables, the high
accuracy of measurements does not really matter. The choice of measurement
accuracy is often arbitrary even when multiple studies involve the
participation of experts from different institutions. In such cases it is
the individual professional orientation and the know how of participants that
determine the choice of pairs for study and the accuracy of measurements for
each of them. What is subsequently revealed in the processing of data thus
collected is, on the one hand, a number of redundancies in some measurements
and, on the other, the paucity of data and insufficient accuracy in others.

CONSTRUCTION OF "STATE INDICES" OF BIOLOGICAL OBJECTS

Consider as an example the following typical situation involved in a study
of multicomponent and multiconnected objects. Traditional analytical
methodology permits their study part by part with the observation of
individual parameters. For instance, plant ontogenesis may be described by
a series of graphs showing the change in time of the mass, length, leaf
surface area, chlorophyll concentration, photosynthesis, respiration
intensity and so forth. The state of an organism at each particular moment
of ontogenesis may be characterized by the numerical value of each single
parameter, simultaneously all or a limited number of "principal" parameters.
Clearly, the first is insufficient, the second (particularly with a large
number of values being measured and a different type of their variation in
time) may bring about contradictory results and the third (seemingly the
best) - the selection of "principal" and "major" parameters - involves,
however, particular difficulties and objections.

Indeed, the concepts "principal" and "major" are always intuitive, and
selection by this principle is arbitrary. Each specialist (a morphologist,
a physiologist or a biochemist) will justifiably apply the term "principal"
to different features of an organism and as a result, the state indices they
will construct will not be identical and furthermore will lack an objectively
common basis for comparison.

The difficulties, arising in constructing the state indices of an object on
an intuitive basis may be illustrated by the perennial and still unsuccessful
search by limnologists for an indicator of a trophic state of lakes. Similar
difficulties are now faced by those specialists in estimating the quality of
natural waters who expect to choose intuitively the "principal" parameters
of the state of drinking water out of many hydrochemical and biological
variables characterizing its quality.

A more rational basis for the elaboration of state indices of a complex
object is provided by an analysis of its internal structure with a view to
revealing its most general properties, expressed numerically. This approach
may be illustrated by the attempts of ecologists to use rank distributions
in communities and ecosystems [6, 13]. The advantages of this approach are
its informality, biological rationale and, at the same time, objectivity in
elaborating state criteria. Its implementation, however, requires a
preliminary knowledge of the object's structure, and the amount of this
knowledge should probably be considerable.

There is another - and currently the most widespread - variety of a systems
representation of an object with the aid of state indices. It requires
neither a preliminary knowledge of the object's structure, nor a preliminary
selection of the "principal" variables but is based on the entire set of data
on the object. Simultaneous measurement of a number of parameters of a
single object is mandatory in this case. The list of parameters may be drawn
up with some special bias or without a predetermined specific orientation.
Synchronously collected data, processed with the use of multivariate
statistics, permit the construction of a small number of artificial variables,
subject to substantive interpreation. Further the object's behaviour and its
individual states are examined in terms of these multidimensional vectors,
serving as state indicators [1, 15]. It should be stressed that the methods
of multivariate statistics are applicable to the description of the
structure and functioning of any complex objects and are therefore justly
considered systems methods.

It was suggested, however, that the elaboration of state indices on the basis
of some multiparameter studies makes as little sense as multiparameter studies
themselves. This view is sometimes expressed by experts on dynamic modelling.
Indeed, the possibility of calculation by mathematical modelling and of
direct measurement by physical modelling of an infinite number of states
obviates the need of describing each particular state. Although this point
of view is well founded in principle, it is unacceptable to many biologists
for a variety of reasons. One of them is the reluctance of specialists to
reduce the number of variables to a minimum (which is required in
deterministic mathematical models) for it considerably simplifies the object
of study. Therefore the description of particular states of a complex object
(without disregarding its structural and functional complexity) remains an
important general biological problem to be resolved by systems rather than
intuitive methods.

REFERENCES

1. P. F. Andrukovich, *The Application of the Major Component Method in Practical Studies.* Moscow, Moscow University Press (1973), in Russian.
2. I. V. Blauberg and E. G. Yudin, *The Evolution and Essence of the Systems Approach.* Moscow, Nauka (1973), in Russian.
3. A. A. Bogdanov, *General Organization Theory, Tectology.* Part I. Moscow (1925), in Russian.
4. V. I. Kremyansky, *Structural Levels of Living Matter.* Moscow, Nauka (1969), in Russian.
5. D. A. Krivolutsky, Systems organization and the theory of evolution, *Zhurnal Obshchei Biologii* 25, No. 3 230-233 (1965), in Russian.
6. A. P. Levich, The extremum principle in the systems theory and the specific structure of communities, in: *Problems of Ecological Monitoring and Modelling of Ecosystems,* Vol. I. Leningrad, Gidrometeoizdat (1978), in Russian.

7. A. A. Malinovsky, The theory of structures and its role in the systems approach, in: *Systems Research Yearbook*. Moscow, Nauka (1970), in Russian.
8. A. A. Malinovsky, Types of biological control systems, in: *Problems of Cybernetics*, Vol. 4. Moscow, Nauka (1960), in Russian.
9. *Mathematical Modelling of Marine Ecological Systems*, Book 2, Leningrad, Leningrad University Press (1977), in Russian.
10. V. N. Sadovsky, *Foundations of the General Systems Theory*. Moscow, Nauka (1974), in Russian.
11. M. I. Setrov, *Organization of Biosystems*. Leningrad, Nauka (1971), in Russian.
12. *Systems Research Yearbook*. Moscow, Nauka (1970), in Russian.
13. V. D. Fedorov, A. P. Levich and E. K. Kondrik, Rank distribution of phytoplankton in the White Sea, *Doklady AN SSSR* 197, No. 1 342-345 (1977), in Russian.
14. K. M. Khailov, The problem of systems organization in theoretical biology, *Zhurnal Obshchei Biologii* 24, No. 5 324-333 (1963), in Russian.
15. V. M. Schmidt, B. R. Vasiliev and S. F. Kolodyazhny, On the generalization of experience in the mathematical analysis of growth and organogenesis in plants, *Zhurnal Obshchei Biologii* 39, No. 3 174-179 (1978), in Russian.
16. E. G. Yudin, *The Systems Approach and the Principle of Activity*. Moscow, Nauka (1978), in Russian.

THE UNDERLYING CONCEPTS OF THE FORMAL THEORY OF WHOLENESS

G. A. Smirnov

Two types of whole objects, i.e. an integral and a whole were introduced by myself earlier [8]. An integral object is an ideal entity no element of which has definitenesses independent of those of other elements. Conversely, an element of a whole object may have definiteness independent of the definitenesses of other elements. Apart from independent definitenesses, all the elements of a whole object must have interdependent definitenesses. These form a relationship of independent definitenesses.

In the above paper [8], I established that a special kind of object, namely integral objects, must be introduced before the problem of defining the ideal whole object can be solved. This is an attempt at a more rigorous analysis of typological characteristics of ideal whole objects. Such an analysis will enable one to state the distinctive features and design philosophy of a formal theory of whole objects. An elementary part of this theory containing a universe of integral objects will be constructed in conclusion.

The problem of defining the type of ideal whole object can be made simpler if a comparison is made with well-known ideal entities like sets. In my previous analysis [8] of the concept of set I concluded that objects like sets are a manifestation of independent definitenesses that has an external foundation for its unity in the form of an object whose function will provide a tie.

I will now try to elucidate the way of unifying a variety into a unity to obtain an object of the type of a set. N. N. Luzin provides the following illustrative example to make it more clear how a set is generated: "Imagine a transparent and impervious envelope like a sealed bag. Assume that all the elements e of a given set M and no other entity are inside. This envelope with entities e inside is an example of a set M made up of elements e. The transparent envelope itself encompassing all these elements (and nothing else) is a fairly good representation of the act of unifying the elements e to obtain the set M ([6], p. 8). This picture gives a rather good idea of the fact that a set is merely generated by uniting or bringing together entities to be elements of the set. "A set is a multiple conceived of as a single" ([6], p. 8). This Kantorian definition of the set by Luzin in a lapidary form states that a set is nothing but a variety isolated somehow from the rest. Or is it?

Any variety consists of entities that cannot all be of the same kind. There are no means within the variety to distinguish an entity of the kind A from another entity of the same kind A.

Any two entities of the same kind A must at least differ by their position in some space or within a sum of available positions. Otherwise they will be mutually undistinguishable. However, entities each characterized by its certain position in a given sum of available positions are not a variety of isolated entities, but rather a unity of this variety, this unity provided as a result of the sum of positions. *No pure variety of isolated entities can contain two identical (in terms of type) entities since this variety has no means of distinguishing between, or correlating these entities.* A variety of entities is made up of unique and noncorrelating definitenesses.

In fact, there are no means within a variety to identify an individual entity. Entities cannot be united into a variety (i.e. into a unity of distinguishable components) if irrelative definitenesses are present in each component only as long as no characteristic of this component points to any other component as being different from the former. *Any component of a variety always has identifying features.* These specify a negative or relative definiteness for each component so that it can be identified and related to other components of the variety. Any component of a variety is a dualism of a positive and a negative definiteness, i.e. has two features, one irrelative of and the other relative to the other components.

Thus a set is not a multiple, i.e. not a variety of objects identified by indicating its boundary. The generation of a set of entities cannot be described as a mere process of combining entities of a variety. No uniting process is possible unless an identification procedure is conducted to specify the relative distinctive features of irrelative definiteness. Using the Luzin image described above, the things in the bag will not represent a single thing (a unity unidentifiable *per se*) but rather the mutually distinguishable components of a given set. This is because they occupy different positions in the total amount of relative positions available within the bag.

The existence of relative distinctive features for components of any set is not solely manifested through the possibility of generating a variety, i.e. a unity of entities whose mutual distinction is assumed as their inherent property. All operations on sets employed to form or transform objects in the theory of sets assume that components have relative distinctive features in any set. We will consider the conjunction of two sets as an example.

A set M_s will be *the sum* of sets M_a and M_b if the set M_s consists of (and only of) elements of the sets M_a and M_b. Let $M_a = \{x, y\}$, $M_b = \{z, w\}$, then $M_s = \{x, y, z, w\}$.

How can we establish that M_s is the sum of M_a and M_b? We single out an element of M_s, such as a positively defined component, z, and relate it to an element of M_a. If the latter is positive, then we proceed by establishing the correspondence between other elements of M_s and that of M_a. If, on the contrary, the element of M_a has a positive definiteness differing from the positive definiteness of z, then we take another element of M_a and match it with the element of M_s having the same definition of z. This procedure does not mean that we do not exclude z_{M_s} (i.e. the element z of M_s) from M_s; rather, we take it as an element retaining both its identity and its relationship with x_{M_s}, y_{M_s} and w_{M_s}.

If x_{M_s}, y_{M_s}, z_{M_s} and w_{M_s} had only irrelative definitenesses, i.e. if they were a delineated variety, the extraction of the object z_{M_s} from the variety as an element of an $M_u = \{z_{M_s}, x_{M_0}\}$ set (this set must be formed so as to match the above elements of the two sets) would result in the destruction of M_s and M_a, since both z_{M_s} and x_{M_0} would then be brought outside their respective boundaries.

On the other hand, a set formed by these elements would consist of irrelative definitenesses z and x, rather than z_{M_s} and x_{M_0}. Specifying this sort of set, $M'_u = \{z, x\}$, would not result in matching elements of M_s and M_a since that which is present in M'_u would be identities irrelative of M_s and M_a rather than elements of these sets. The fact is that matching elements of sets, i.e., establishing a correspondence between positive definitenesses of these sets, does not result in the destruction of the matched sets. This fact shows that instead of obtaining simply an apparent union of positive definitenesses, we have a conjunction obtained by imparting relative features to positive definitenesses. These relative features do not disappear, whatever might be the way these elements are matched. The set $M_u = \{z_{M_s}, x_{M_a}\}$ contains elements z_{M_s} and x_{M_a} with mutually irrelative definitenesses. Both z_{M_s} and x_{M_a} retain their relationship to other elements, although they become elements of a new set, M_u. This is the only reason why M_u functions as a set that matches elements of M_s and M_a. The possibility of forming a set out of *elements* from different sets (rather than just out of positive definitenesses characterizing these elements) reveals the inherent correlation of the elements within a set, i.e. the fact that each element has a feature relating it to the other elements of the set.

Any relating (matching) feature is a *distinguishing* feature and not only because relating features permit the distinguishing of elements with the same irrelative definiteness. To match elements from different sets, say M_a and M_s, we must match each element of M_a with each element of M_s. After matching x_{M_s} and x_{M_a} we must match elements of M_s *other than* x_{M_s} with elements of M_a *other than* x_{M_a}. We must go to elements of M_s *other than* x_{M_s}. This transition assumes that x has a feature distinguishing this element from the other elements (and only from these) of M.

Thus a set is a unity of irrelative definitenesses, i.e. such that is provided by matching these with correlating definitenesses. Any element is characterized by a negative definiteness, i.e. by features distinguishing it from the other elements of an object.

Distinctive features form a *conjunction* of irrelative definitenesses. This conjunction is a prerequisite for generating and manipulating objects in the set theory. This prerequisite manifests itself in distinguishing previously undistinguished objects as different elements of a set, and in the

possibility of matching elements from different sets. No set can be formed
from a variety of irrelative definitenesses unless a conjunction of
distinctive features is present to establish relative definiteness between
elements of the set. No distinctive features appear if a number of objects
are extracted from the variety by some extraneous delineation. Any relating
feature must be an entity independent of the variety if this feature is not
contained therein. If a set can be generated from entities of a *variety*,
something independent of this variety must be available to be a basis of the
unity.

*The generation of an object assumes that the generating entities are mutually
independent.* If only interdependent definitenesses can be identified in an
object, these cannot be regarded as its constituent entities and the object
is therefore simple (indivisible). A set is a compound object generated
from a variety, since irrelative definitenesses present in any set are given
independently of the set and as a condition of its generation.

In set theory, the extensionality concept is adopted. According to this
principle any set is defined by its components, or more precisely, by the
irrelative definitenesses of its constituents. Sets as seen in the light of
this theory are objects made up of uncorrelating definitenesses. Specifically,
sets may be composed of components of different sets.

*Either we assume a set to be a compound object generated from a variety of
noncorrelating entities - which means that a basic principle independent of
the variety is required to obtain a set; or we must consider a set as a
simple and indivisible entity.* An object in the set theory is assumed to be
compound and formed by elements of a variety. Then the assumption of a
conjunction defined independently of the variety is a necessary condition
for generating a set.

The presence of conjunction as an object constituted by negative
definitenesses becomes apparent in set-generating events. I have shown
above that the presence of conjunction is a necessary feature of a set, as
a basis for the unity of a generated set becomes apparent in the way the
objects of the set theory are manipulated. However, the assumption of
conjunction - although a prerequisite of the set theory - is neither
formulated nor analysed within the framework of this theory. The latter
confines itself to studying correspondences between positive definitenesses
of elements from different sets while neglecting the problem of what makes
this comparison possible as well as the problem of a prerequisite for
generating set-like objects.

The definition of a whole ideal object is therefore the problem of defining
an object "conjunction", i.e. the grounds for the integrity of a compound
object. Conjunction is a single object containing correlating definitenesses,
i.e. distinctive features. A whole object of any type contains negative
definitenesses among the definitenesses of its elements. Now let us proceed
to a concrete discussion of whether the assumption of this object is actually
possible.

Any element of an integral object must possess only a definiteness which is
not independent of the definiteness of another element. The elements of an
integral object must be correlating (i.e. they must assume mutual existence)
rather than irrelative, independent definitenesses. My previous analysis of
Plato's *Parmenidus* [8] convinced me that an object disintegrates into a
variety of independent entities if at least two concurrent definitenesses can
be identified as constituent elements of this object. Indeed, if a

definiteness A exists concurrently with a definiteness B, the existence of A
does not imply that of B. B might be a prerequisite for the formation of A.
However, having appeared, the definiteness A does not need B as a condition
of the existence of A. The definiteness B will be a prerequisite of the
existence of A and A will assume B if the assumption of A is a result of
previous assumption of B. A and B will assume each other if assumption of
one definiteness is a result of the earlier assumption of the other. Only if
one of the two definitenesses (A) disappears when it is replaced by the other
(B) is the existence of A a prerequisite for the existence of B. A and B
will have correlative or negative definiteness if the assumption of
definiteness forestalls the assumption of B (and vice versa): A and B
themselves should exist only in the course of turning into each other. A and
B should be definitenesses which are established in the process of being
distinguished and do not exist outside that process. The constitutive
distinguishing procedure that involves the interchange of the definitenesses
assumed in this procedure is a necessary condition of generating an integral
object. However, any element as an item of this procedure has not an
irrelative definiteness, but rather a definiteness pointing to that of other
elements, i.e. correlates with them.

A whole ideal object of whatever type is only possible as a process of
constitutive element distinction whereby no two elements can coexist. Not
only can elements have relative features, but the very decomposition of a
whole into elements is made possible due to these relative features.
Elements of a whole are definitenesses appearing through the disappearances
of each other.

Elements of an integral object, like elements of any wholeness, are thus
characterized not only by *what* they are (the essential feature) but also by
how they exist (the existential feature). One cannot define what an element
A is unless one indicates the element B wherein A disappears; since the
definiteness A exists only in the distinguishing process, the essential
definiteness of A is inseparable from the existential definiteness of this
element. Only for irrelative definitenesses can one separate what they are
from the sequence of events assuming them. This sequence is not necessary
to assume those definitenesses, but is rather an externally given sequence
of mutually independent events. One irrelative definiteness (A) can be
given independently of another (B), hence A and B can either coexist or
follow each other. The interrelation between A and B will be determined by
something else, i.e. by some other definiteness C that will specify an
existential characteristic, or a way to correlatively assume A and B with
mutually independent definitenesses.

Correlative elements must thus be established through a distinguishing
procedure inseparable from their mutual assumption process. Why should we
consider the distinguishing procedure, a purely logical one, as an existential
feature of an integral object?

A whole ideal object must be such a unity of a variety that has its foundation
in itself. Both the unity and distinctions derived within a whole ideal
object must be present within this very object, or else we will have an
externally imposed unity of a variety (a set) instead of an internally imposed
variety), a wholeness. If the unity of set-theoretical objects and hence the
relative distinctive features of elements, only appears in the course of
distinguishing and correlating procedures carried out by a subject while
manipulating these objects, and is not contained in the objects themselves,
then a *basis of this unity must be postulated as a feature of a whole object,
necessary to generate the latter. The distinguishing procedure must be an*

objective procedure of assuming correlating distinctive features rather than a logical, i.e. subjective procedure. The former is the object itself.
Unlike a set, a variety of positive definitenesses that are given in a mutually independent manner and are unified by an external procedure, a whole object is a procedure of constituting distinctive features that cannot exist outside this procedure. The variety of distinguished elements will then be a result of this procedure which will at the same time be a basis for the unity of this variety, while no external technique will be required to unify the distinctive features.

A logical definition of the type of an integral object as a wholeness containing only mutually assuming definitenesses as features of its elements becomes an objective definition, a procedure for establishing elements, if this definition indicates the way those elements live, or the way they are specified. If a whole object consists of two elements, the logical assertion "A and B mutually assume each other" has its correspondence in the procedure of the appearance and disappearance of definitenesses as a process of constituting A and B whereby A appears due to the disappearance of B (and vice versa); A and B negate each other, one definiteness is assumed to be distinct from the other, or to be a non-other; however, one of the elements is not externally distinct from the other, it rather undergoes a distinguishing process; hence A disappears and B appears. In this process of appearance and disappearance, A and B both negate and assume each other.

Since A and B are themselves involved in an objective correlating process – that is in a process of the appearance and disappearance of definiteness as steps in a constitutive distinguishing process – these A and B can afterwards be logically correlated.

To develop a theory of ideal whole objects, one must include, as features of an object universe, all the prerequisites for the logical procedures that can be conducted on these objects. A logical procedure is a way of correlating. If definitenesses are correlated externally and the correlating procedure is considered to fall outside the framework of the object universe available in the theory, the latter will be a theory of definitenesses externally unified and given in a mutually independent way, and not a theory of whole objects. The object universe in a theory of whole objects must be complete; in a whole object, both the relating item (a tie) and the related items (independent definitenesses) must equally belong to the universe itself.

We consider a formal theory of the whole object as providing rules to combine simple constitutive distinguishing procedures into more complex ones. The result of any such combination must be a distinguishing procedure constituting its own elements. More complex procedures should include various types of subordination, i.e. a hierarchy of simpler procedures. A question naturally arises: what can be considered an empirical interpretation of these constitutive procedures? Prior to formally constructing the simplest piece of this theory, realistic formations have to be indicated that represent at least the underlying ideal objects of the theory.

Any complete description of the set theory begins with the presentation of examples to illustrate a basic intuitive set, such as a pile of rocks and a flock of birds (e.g. see [1], p. 13 and [6], p. 7). These by no means imply that birds and/or rocks will be treated by the mathematical theory; those entities are only intended to point out tangibles that are or might be ascribed to any individual and have immediate validity, that is they are specified so as to preset an underlying intuition of the ideal object whose analysis and development are the goal of the theory.

What realistic formations can be indicated as an empirical example of the simplest integral object, i.e. a procedure to constitutively distinguish two elements (a dyad)? It is clear that no dyad intuition can be specified by indicating an object given in sensual perception. We always have in sensual perception a variety consisting of individual and mutually independent definitenesses. We can, for example, perceive two actual definitenesses A and B as either coexisting or sequentially existing: this perception involves perception of A, perception of B and perception of their spatial and time relationship. The necessity of having A for B to appear – that is, the availability of A and B as steps in a common process – is not an object to be perceived. In perception, any sequence or coexistence, i.e. any correlating procedure, acts as a spatial and time correlation that does not establish correlation stages but rather coassumes some definitenesses existing by themselves. No science would be necessary if the necessity to correlate were an object to perceive.

The dyad, or the simplest constitutive distinguishing procedure, can be represented in life events rather than in perceptions. Man is not only a cognizing and perceiving creature, he is above all characterized by his changing inherent states. Breathing and movement, that is, the processes characterizing a human being as a living creature, are examples of such states. A person is alive as long as he breathes. Breathing is the change from inhalation to exhalation, that is, change between two states, neither of which can exist without the other. Correlation between inhalation and exhalation as stages of a single process is realized by every human being to be valid as the distinction between these stages. Similarly, a person's movement is directly realized change between the correlating stages of muscular tension and relaxation. Breathing can be regarded as a real formation representing a dyade: inhalation both excludes and assumes exhalation (and vice versa), with the appearance of one of these elements being possible only provided that the other disappears. Neither inhalation nor exhalation can exist outside the breathing process, breathing constituting inhalation and exhalation as its phases.

In this approach, breathing as the change between inhalation and exhalation or movement as the change between tension and relaxation are treated as primary constitutive procedures; human life activity then appears to be a dualism of these procedures or an n-ary procedure involving breathing and muscular processes as well as many other representations of a binary distinguishing procedure.

Other actual whole formations can be described in a similar way as complex distinguishing procedures consisting of dyads. I shall confine this discussion to the above since my objective was to give an example of a simple and directly valid whole object, that is, to give a basic intuition for a well-established theory of wholeness and the type of objects in this theory and that objective has been achieved. The further refinement of this intuition should be based on a formalism that makes explicit the techniques to combine simple procedures into more complex ones.

In this respect I note that the understanding of an ideal whole object as a constitutive distinguishing procedure, as well as the idea that a theory of the whole object must involve all the prerequisites for generating and transforming objects as characteristics of a universe itself, have their roots in a certain logical, philosophical and scientific tradition. The theory of opposites that underly every existing entity has been a key

element of any dialectical system since antiquity. It is well known that
this theory is not limited to existing entities, this means that it is not
just ontology; in fact, it determines a technique for obtaining ideal objects
and a logic to be used in theoretical reasoning. I will consider some
logical aspects of this theory in order to compare them to the characteristics
of ideal objects in a formal theory of wholeness. Of course, I do not claim
to be able to present in this logical and mathematical theory all the wealth
of the dialectical theory of the unity and struggle of opposites, which is
most comprehensively presented in the philosophy of dialectical materialism.
I wish only to point out the similarity and difference in formal constructs
between the theory of wholeness and those dialectical systems in which
special emphasis was given to the description of the logical aspect of the
theory of opposites, namely the Platonian and Hegelian systems.

The basic intuition in these systems is not a concept of individual
definitenesses but rather that of pairs of definitenesses (opposites), such
that one of the definitenesses is preconditioned by the other in any one
pair. Two different understandings can be identified. One treats opposites
as mutually different definitenesses, and each definiteness correlates to,
rather than contains the other. This is Plato's view: "An opposite itself
can never become its own opposite" (*Phaedo*, 103b) [7, pp. 148-149], from
which it is clear that "...no opposite will ever be its own opposite..."
(*Phaedo*, 103c) [7, p. 149]. Hegel takes a different position. In any
concept, Hegel's *Science of Logic* maintains it must be "demonstrate... that,
being what it is, it is its own opposite, that its opposite has penetrated
into it as being what it is, that it itself is the perfected transition from
itself, namely, determinateness" [4, Vol. 1, p. 111]. Each definiteness
contains its own negation specifying the transition from this definiteness
to another. "That by means of which the concept forges ahead is the...
negative which it carries within itself" [4, Vol. 1, p. 66].

The existence of opposites is not the only thing Hegel insists on. He claims
that the opposite is present in any definiteness: "something is itself and
is also deficiency or the negative of itself, in one and the same respect"
[4, Vol. 2, p. 68].

Since the definition of a binary distinguishing procedure specifies a fixed
distinction between correlating definitenesses, one might see in that
definition an analogy with Plato's understanding of opposites. However, my
definition of a binary distinguishing procedure differs from that of Plato
in another respect, namely in that the relationship between these
definitenesses is understood as an existential characteristic of opposite
definitenesses. Plato asks the question, through Socrates' mouth: "Whether
it is non-necessary that, when a thing has an opposite, it can come to be
only from that opposite" (*Phaedo*, 70e) [7, p. 60]. He appears to answer
positively: "the actual occurrence must conform to our principle: they all
come into being from each other, and there is a process in which each
becomes the other" (*Phaedo*, 71b) [7, p. 60]. However, for Plato this answer
applies only to opposite things rather than to the opposites as such. Only
that which has a relation to the definitenesses rather than the definitenesses
as such can convert one into the other. Plato related the definiteness of
distinctions, like any other definiteness, to the absence of any change
whatsoever, and according to him, any indefiniteness has its origin in
variability.

Plato thinks that the only way to maintain a definiteness is through its
delineation, independence or separation from any other definiteness. For any
separate and independent definiteness, any change implies that the

definiteness will disappear and degenerate into another one.

While leaving aside the analysis of the philosophical and gnoseological roots of the above, I would only like to point out its purely logical inconsistency. Plato bases his view on the idea that any change is indefinite and contains no regularity. With that kind of change assumed, one can indeed retain a definiteness, but only by assuming invariant and independent entities. This is the guideline that Plato follows and this way of retaining a definiteness is characteristic of mathematics. We must deal with clearly distinct definitenesses if we are to build a mathematically rigorous and clear theory, and we cannot deal, in the framework of this theory, with inconsistent definitenesses, since inherently these do not have clearly assumed distinctions.

However, in addition to entities understood as objects of a universe in a mathematical theory, such a theory defines procedures to transform objects, to change them in a regular fashion. The purpose of these procedures is to enable the handling of objects either fixed in the language of the theory, or assumed as a condition for performing certain operations. We have shown that these object-transforming procedures are based on distinguishing procedures, that is, on the pre-establishment of correlative distinctions.

It is invariance of the distinguishing process rather than that of the independent definitenesses (i.e. the results of a distinguishing procedure) that can form a basis for assuming clearly outlined distinctions. If the procedure of change is primary in respect to its definiteness, that is, if distinctions are constituted by the procedure itself, the procedure invariance is a basis for correlative distinctions being clearly definite. If, to the contrary, the process of change is secondary and the changing entity is given from outside this process, change will amount to eliminating the definiteness involved. Either we will, like Plato, abandon the assumption of change and then a universe of ideal objects will be a basically unintegrable variety of invariant definitenesses. Or we shall postulate, as primary objects of our universe, regular procedures constituting definitenesses.

It was noted above that the construction of this sort of universe implies that this universe includes all the prerequisites for producing whole objects. In set theory, what *is* correlated is the object, and what correlates is a correlating procedure performed by the subject and is partially registered in that theory's language. In contrast, no such separation is or can be present in a formal theory of wholeness. Either it can involve a separate procedure with correlating elements or a procedure can act in this theory to correlate other procedures within the framework of a single complex procedure that establishes the resolution into elements. Whatever the case, both the procedure underlying correlation and the procedures to be correlated belong to an object universe and are parts of a single procedure that constitutes correlating elements (i.e. definitenesses) existing only within the framework of the procedure.

Ideally whole objects can be presented in the framework of such a theory, whereby neither production nor transformation of objects demands the existence of a subject or the presence of some subjective capabilities. I have shown that the set theory tacitly implies the existence of a subject that correlates and distinguishes elements of a set as well as elements of different sets.

It is generally accepted that the most detailed theoretical reasoning in favour of a concept of a scientific theory such as a universe of ideal objects whose production and functioning depends on the subject's activity was provided by Kantian philosophy. The necessity of abandoning the understanding of the subject as an entity initiating distinctions within the substance of theory was clearly recognized by Hegel. In his *Phenomenologie des Geistes*, he contrasts discursive thinking present outside the subject that compartmentalizes thinking externally, with thinking in terms of concepts, as the only possible way to assume a wholeness.

> "...just as ratiocinative thinking in its negative reference ... is nothing but the self into which the content returns; in the same way, on the other hand, in its positive cognitive process the self is an ideally presented subject to which the content is related as an accident and predicate. This subject constitutes the basis to which the content is attached and on which the process moves to and for. Conceptual thinking goes on in quite a different way ... It is not a quiescent subject, passively supporting accidents: it is a self-determining active concept which takes up its determinations and makes them its own. In the course of this process that inert passive subject really disappears; it enters into the different constituents and pervades the content; instead of remaining in inert antithesis to determinateness of content, it constitutes, in fact, that very specificity, i.e. the content as differentiated along with the process of bringing this about [3, pp. 118-119].

The introduction of distinguishing procedures as primary objects in a formal theory of wholeness rests upon a logical and philosophical tradition as well as on the treatment of the problem of wholeness within the framework of systems research. In their examination of ways to resolve wholeness and of the relationship between elements within a whole, Blauberg and Yudin write: "The experience gained in the course of complex object research shows that instead of a simple functional relationship, a far more complex set of ties exists between parts of a complex object (as well as between parts and the whole).

Within this set of ties, the cause is at the same time the result of these ties and this result is assumed to be a prerequisite. In other words, the interdependence of the parts is such that it manifests itself not as a linear causal sequence, but rather as a peculiar closed circle within which each element of a tie is a condition for another one to exist and is stipulated by the latter. This was pointed out by Karl Marx in his analysis of the system of bourgeois economic relationships and extended to encompass all organic systems [see: K. Marx, *Grundrisse der Kritik der politische Ökonomie* (Rohentwurf). Berlin, 1953, S. 189] ([2], p. 125).

We see that the systems approach to the study of a complex object follows Marx in considering the presence of a cyclic procedure with mutually assuming elements of a whole to be a typical feature of a whole formation.

Blauberg and Yudin point out another necessary attribute of a whole object. Systems research of complex objects involves singling out a structure that represents the composition of such objects. Blauberg and Yudin note that the composition of a whole object cannot be found unless one examines the functional, or the existence dynamics, of the object involved. "The structure is not a dead replica of a frozen object but rather a characteristic of those of its aspects which are highlighted only in the course of analysis of its actual dynamics. Therefore the structural characteristic is, among other things, a dynamic characteristic" ([2], p. 139). This makes it "necessary to resort to time-domain characteristics, thereby making time a compulsory component of any structural and functional description. Yet this is evidently not the historical time one deals with when analysing the evolution of the object, but rather a special type of time which might be called functioning time" ([2], p. 139).

The assumption of an ideal whole object as an element-constituting procedure makes it possible to pinpoint in an ideal object the most significant sign of wholeness, namely the fact that elements of a whole object are presented in the process or the dynamics of their mutually stipulated existence. From the outset we must introduce a dynamic characteristic in an ideal object, and present the object as a certain process or procedure if we are to obtain an ideal whole formation.

I will not go into detail on the analysis of the methods of assuming an ideal object in various sciences. I only point out that in many sciences the distinguishing procedure plays the role of the procedure of assuming correlative definitenesses. This fact is most explicit in linguistics and semiotics in which fundamental objects are chosen on the basis of a binary criterion (see, for example, the paper by Ivanov [5]).

I conclude this chapter with a formal description of a universe of single (integral) objects. For this purpose, I will introduce a notation representing certain characteristics of the universe, objects contained therein and methods of defining these objects. The letter D with subscripts will record the formal introduction of a combination of symbols referring to a universe from the theory of whole objects.

D_1. U_e is a universe of integral objects. This universe includes universes of primary object (or dyads) and the universe of derivative objects (or aenads). D_2. U_d is the dyad universe and D_3. U_E is the aenad universe.

Any object of U is of the integral object type, and is a wholeness of n elements, each element possessing a definiteness correlating to the definitenesses of the other elements.

D_4. $[E_{i1}] \leftrightarrow \ldots \leftrightarrow [E_{in}]$ (denoted in abridged form as γ_n) is the universe object containing $[E_{i1}], \ldots, [E_{in}]$ as elements where E_{ij} is the element definiteness.

Given these introductory definitions, we will now describe U_d and U_E.

(1) The universe of dyads, U_d.

The universe of dyads U_d includes principles to generate (define) objects, the defining procedure, and results of this definition, i.e. dyad objects.

D_5. Ω_1 is the starting point of binary distinguishing procedure; D_6. $\{S_i\}$ is a conglomerate of underfined substrates.

It is necessary to introduce Ω_1 and $\{S_1\}$ since one cannot just postulate a variety of differing objects, each being a procedure of distinguishing two objects (a decomposition into two elements). According to the construction criteria permitting a theory to include all the prerequisites, the presence

Formal Theory of Wholeness

of both the difference and identity of primary objects requires that principles underlying identity and difference be postulated.

D_7. $\Omega_1(S_i) = [\alpha_{i1}]_1 \leftrightarrow [\alpha_{i2}]_2$ is a dyad-defining procedure (a procedure to match Ω^1 and S_i principles).

D_8. The dyad $[\alpha_{i1}]_1 \leftrightarrow [\alpha_{i2}]_2$ will be denoted simply as Δ_i.

As a result of D_7, a certain object is produced which is of type D_1, implying that it is composed of elements.

The dyad $[\alpha_{i1}]_1 \leftrightarrow [\alpha_{i2}]_2$ contains two elements with correlative definitenesses. It should be emphasized that since S_i is an indefinite substrate, only Δ_i is a definite object and neither α_{i1} nor α_{12} is a result of decomposing the already available S_i definiteness into two parts. The definitenesses α_{i1} and α_{i2} appears inside Δ_i while the presence of two different substrates results in the difference between Δ_i and Δ_j.

(2) The universe of aenads, U_E.

A new feature pertaining to the U_E universe is that here definite distinguishing procedures play the role of object-defining principles.

The entire variety of objects that can be defined in U_E breaks into aenad classes each containing aenads of the same level different from that of any other class.

D_9. $\{\gamma_k\}$ is a class of k-level aenads.

D_{10}. $\{\Delta_i\} = \{\gamma_0\}$. All the objects from the dyad universe are included in the zero class of objects from the aenad universe.

D_{11}. Ω_2 is the starting point of matching the distinguishing procedures that is the starting point of the defining procedure.

The starting point Ω_2 chooses an object $\gamma_k = [E_{i1}]_1 \leftrightarrow \ldots \leftrightarrow [E_{ik}]_k$ out of $\{\gamma_k\}$ as the basic principle, an element $[E_{ij}]$ out of γ_k as the constitutive element and one of the dyads, $\Delta_n = [\alpha_{n1}]_1 \leftrightarrow [\alpha_{n2}]_2$ as a subsidiary principle. Then Ω_2 matches these two principles.

D_{12}. $\Omega_2([E_{i1}]_1 \leftrightarrow \ldots \leftrightarrow [E_{ik}]_k, [E]_{i\ell}]_\ell, [\alpha_{n1}]_1 \leftrightarrow [\alpha_{n2}]_2) = [E_{i1}]_1 \leftrightarrow \ldots \leftrightarrow [E_{i\ell}\alpha_{n1}]_\ell \leftrightarrow [E_{i\ell}\alpha_{n2}]_{\ell+1} \leftrightarrow \ldots \leftrightarrow [E_{ik}]_{k+1}$ is the procedure distinguishing an object, γ_{k+1}, from the object class whose level follows that of γ_k.

The result of D_{12} is a procedure of identifying γ_{k+1}. This procedure includes a hierarchy of γ_k and Δ_n procedures.

In γ_{k+1}, the existence of the γ_k procedure provides for the existence of the Δ_n procedure due to the following: now Δ_n can only exist if the element

$[E_{i1}]$ is present and cannot exist by itself. The existence of the procedure element $[E_{i1}]$ leads to the existence of the Δ_n procedure in general, while the disappearance of $[E_{i1}]$ results in the disappearance of Δ_n.

Once the existence of a procedure (γ_k) is a prerequisite for the existence of another (Δ_n), a full hierarchy of procedures is available; this is a hierarchy of existential features of whole objects. The concept of the existential feature of an object on the whole is now introduced. The γ_i procedure on the whole can exist either independently or provided that an element exists from the γ_j procedure.

γ_{k+1} is not a mere statement of a hierarchy of γ_k and Δ_n procedures. In fact, in this hierarchy Δ_n is a procedure identifying the definiteness of the $[E_{i1}]$ element. A decomposition into elements is given in γ_{k+1} such that the coexistence of E_{i1} and α_{n1} plays the role of an element different from the element with the definiteness $[E_{ij} \alpha_{n2}]$. The definiteness E_{ij} present in one of these elements is different from the same definiteness in another element; in the successive existence of γ_{k+1} elements the definiteness E_{i1} twice appears and twice disappears as the definitenesses of two elements. The Δ_n procedure establishes a distinction within E_{i1} so that the discriminated element E_{i1} does not coexist.

The identification procedure for γ_{k+1} based on the Ω_2 principle not only specifies a procedural hierarchy but, in fact, determines *how* the elements of the procedures to be matched will coexist as identities of elements of the new procedure.

With a different technique of element matching, the identity of an element, E_{i1}, in the γ_k procedure may be specified by Δ_n on the whole instead of being identified by Δ_n elements. The transition from α_{n1} to α_{n2} and back again will not then be accompanied by the appearance and disappearance, respectively, of E_i but will rather occur during the existence of E_{i1}. The correlative definiteness of E_{i1} acquires a feature that does not relate to any other element of γ_k. The matching procedure will contain an element $[E_{i1}\Delta_n]$ wherein the Δ_n procedure secondary to the E_{i1} identity will specify one and the same element. The existence of the E_{i1} definiteness will then be a reason for the identity of this element. This technique of specifying matching results in a whole rather than an integral object.

We have described the construction of a whole object different from an integral object in order to explain the technique of assuming either a hierarchy of the very procedures, or various types of matching of elements from the procedures matched as the definitenesses of elements of the procedure to be specified. The aim of this paper was confined to the description of a simple element of a formal theory of wholeness so that this element contains a universe of integral objects.

I will attempt to show elsewhere that on the basis of the above piece of theory universes of objects can be specified with the objects corresponding to those of the logic of utterances and the logic of predicates, which are the theories forming the basis of modern mathematical sciences.

This comparison will make it possible to evaluate the formal theory of wholeness as a mathematical tool rather than simply a theory aimed at solving a set of problems involved in constructing an ideal whole object.

REFERENCES

1. P. S. Aleksandrov, *Introduction to a General Theory of Sets and Functions*. Moscow, Leningrad (1948), in Russian.
2. I. V. Blauberg and E. G. Yudin, *Development and Essence of Systems Approach*. Moscow (1973), in Russian.
3. G. W. F. Hegel, *The Phenomenology of Mind*. London, Macmillan (1931).
4. *Hegel's Science of Logic*, Vol. 1 and 2. London, George Allen and Unwin (1929).
5. Vyacheslav V. Ivanov, Binary structures in semiotic systems, in: *Systems Research Yearbook 1972*. Moscow (1972), in Russian.
6. N. N. Luzin, *Theory of Functions of Real Variables*. Moscow (1948), in Russian.
7. *Plato's Phaedo*, Cambridge, Cambridge University Press (1955).
8. G. A. Smirnov, On a definition of an ideal whole object, in: *Systems Research Yearbook 1977*. Moscow (1978), in Russian.

Modelling Methods in Systems Analysis

NONFORMALIZED COMPONENTS OF MODELLING SYSTEMS

N. I. Lapin

The fundamental role of conceptual and, more generally, nonformalized components in a system of global modelling is now gaining prominence as more experience in constructing models of global development is acquired and these models are experimentally studied. This was clearly demonstrated at two international workshops held in the second half of 1978 — the Seminar on Large Systems and Global Models (Dubrovnik, Yugoslavia) and the Sixth Conference on Global Modelling sponsored by the International Institute of Applied Systems Analysis (Wesendorf, Austria). Many participants pointed out that the efficiency of a global development model depends on the conceptual guidelines adopted by the model designer, and primarily on the socio-economic and socio-political concepts [25, 26, 29].

In tackling problems of global modelling, Gvishiani, Zagladin, Frolov and other Soviet scholars, guided by the theory and methodology of Marxism-Leninism, have from the very outset concentrated on the ideological, philosophical and socio-political aspects of this new area of research [5, 6, 11, 12]. This led to the possibility of a conceptually different, broader and deeper approach to the very process of formulating goals and targets for global modelling seen as a tool for studying development alternatives for the world or its regions. On this basis several special methodological problems associated with the construction of a modelling system have been identified and explored, while sociological [13, 14, 17] and applied (e.g. [23]) problems of global modelling have also been studied.

The construction of a modelling system of global development is a comprehensive exercise combining many disciplines and demands the use of systems-theoretical tools. It involves several methodological problems, among them that of the general topology of nonformalized components of a modelling system, as well as the more specific problems of constructing the conceptual components of such a system. These two types of problems will be considered below.

GENERAL TOPOLOGY OF NONFORMALIZED COMPONENTS

Soviet authors treat the modelling system as a man-machine system involving the use of both formal and nonformal techniques. The formal components were described by Gelovani and Dubovsky [7, 8], while the nonformal components were defined by myself [15]. A generalized approach to the topology of nonformalized components of a global modelling system is discussed in the sequel. I chose not to get involved in any dispute over the definition of a "model"[*] and only assume that any modelling system (MS) is a man-machine system. It contains a set of formalized and/or nonformalized data about the model object that is required to treat problems which will be posed by the analyst (the user). (The solution of these problems is the goal of constructing the MS.) This set of data is ordered according to the structure of the model object and the laws governing object functioning and evolution. A data component described in a formal language such as mathematical or algorithmic will be referred to as a formalized component, while a data component described in a natural language will be called a nonformalized component. Of course, the distinction between these two types of components is only relative, since a component nonformalized in one MS may be formalized in another. In other words, the borderline between the two types is determined by the MS goals, the mathematical formalism used, etc. However, this borderline can be defined precisely and unambiguously for any particular MS. I will classify nonformalized components according to three types: nonformalizable, formalizable and deformalizable (Fig. 1).

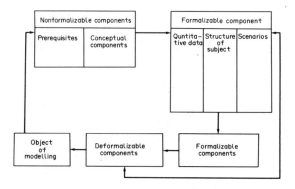

Fig. 1. Block diagram of modelling system components.

The *nonformalizable components* initially include the prerequisites to construct an MS, especially the underlying information about the object to be modelled, and in particular, the information on both actual and potential problems of object evolution. A tentative pragmatic list of problems derived from this information is the starting point of analysis. With this list, one can further elaborate the goals and targets as well as decide on the procedures to be used in the course of construction. The following systems-theoretical principle should be borne in mind: one must avoid confusion between problem formulation and solution procedures so that the former need not be changed in case the latter is revised [20]. This implies that the future MS must as a whole be invariant, although some of its components may be relatively variable.

---

[*]One of the best definitions of a "model" appears that given in [18].

Second, most conceptual components of an MS are nonformalizable. These include primarily the basic theoretical and methodological principles of the concepts taken as the underlying philosophy of the entire modelling system. From these principles, an initial object structure is derived. This structure must contain a minimum of major structural components, i.e. the set of these components must provide an adequately structured empirical list of object problems. Further it is necessary to develop, on the one hand, a detailed structure of the object and, on the other, a set of problems and object evolution alternatives, including generalized alternatives. A specific concept must be designed before the two procedures can be implemented. This concept must make use of the principles of the basic concept and be an euristic factor throughout the rest of the construction procedure. This means that the specific concept must be aimed at achieving concrete goals and finding special techniques to implement a particular MS. A block diagram of conceptual components is shown in Fig. 2.

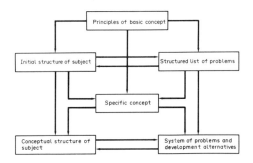

Fig. 2. Conceptual components of a modelling system.

The *formalizable components* serve to relate the nonformalized and formalized parts of an MS. This happens because the formalizable components are essentially written in terms that can be translated into formal languages. First, these terms include the formalizable object structure containing only the conceptual structure components to be formalized. Second, they include nonformalizable information concerning object processes. The requirements as to the content of this information are determined by the formalizable object structure. However, the lack of appropriate information imposes rigid limits on this structure. Third, the formalizable MS components include the scenarios enabling analysts or other MS users to set, study (via the formalized part), and assess alternative evolution patterns of the object (i.e. the world or its regions). The range of scenario plots is limited by the capabilities of the formalized MS part. It is by no means accidental that even the most flexible in terms of their scenarios global models of Mesarovic-Pestel [28] and Leontiev et al. [27] can only be used to study economic and resource problems. On the other hand no adequate formalism has yet been found for the set of multipurpose scenarios developed at the University of Sussex, U.K. by a team headed by Cole and including socio-economic and socio-political problems [30]. The construction of an MS that allows rather extensive use of modern computer capabilities in studying multipurpose scenarios is one of the most pressing problems of global modelling.

The issue of *deformalizable MS components*, i.e. of procedures to interpret MS construction results in a qualitative and nonformal way, has not been adequately treated. An attractive idea has been suggested of introducing a computer interpretation unit in the MS so that the results can be rapidly analysed and presented as verbal descriptions of some typical situations [7]. Procedures are also required to provide adequate "up-conversion" of data obtained from the formalized MS part to conceptually interpreted results. The analysis of these results may require the modification of some formalizable components (e.g. a scenario) and/or nonformalizable components (e.g. a set of evolution alternatives). This implies that the nonformalized MS components must be sufficiently flexible and modifiable.

Now I turn to more specific problems involved in the construction of conceptual components for a global evolution modelling system.

CHOICE OF BASIC CONCEPT

The problem of basic concept is vital to the construction of a global evolution MS as a whole, and especially to the elaboration of conceptual components of this system. The entire global modelling project can be affected if this problem is neglected or underestimated.

The global nature of the object of modelling imposes specific requirements on the basic MS concept:

> ideological comprehensiveness and definiteness so that the most significant structural and dynamic behaviour of an individual or a society of humanity or nature, can be treated from a common standpoint;

> historical approach as a concept fundamental to establishing the evolutionary trends of the world or its regions, combined with the aim of disclosing the inherent sources of this evolution;

> receptiveness to fresh scientific results and the ability to find efficient ways of handling difficulties; and

> a strictly defined conceptual apparatus so that the theoretical substance of the basic concept can largely be formalized.

Global modelling in the West lacks this sort of comprehensive basic concept. The authors of certain Western models use, whether deliberately or not, widely differing and possibly contradictory elements of philosophical, sociologic, economic and political concepts. Marxist scholars, on the contrary, have conceptual advantages in designing a global evolution MS as it is Marxism-Leninism that provides a consistent theory on whose basis this challenging problem can eventually be solved.

First, Marxism is a truly scientific ideology which has evolved through scientific generalization and understanding from the position of the leading social class, of human existence as a whole in its various manifestations, rather than of an isolated aspect of human activity [22]. This ideology is unique in its comprehensiveness and multifaceted nature and encompasses all basic realms of existence and consciousness of the individual human being, society and humanity. The philosophy of Marxism integrates the fundamental achievements of all the arts and sciences. The sociology of Marxism defines a set of concepts representing laws of social progress, the structure and the dynamics of various types of societies. The economic theory of Marxism

Nonformalized Components of Modelling Systems

analyses the laws of economic evolution of the human race, regions or individual countries and discloses the basis for socio-political transformations. The political theory of Marxism reveals laws of class struggle and revolutions, as well as the functions, structure and forms of political power. The theory of socialism and communism is a comprehensive (i.e. philosophic, sociologic, economic and political) substantiation of laws, forms and ways of building socialism and communism as the highest stage of social evolution.

Progress towards this stage is the main trend of the modern and coming epoch of human history.

Second, all the constituent components of Marxism-Leninism, although they do have specific features pertaining to particular objects and functions, complement and affect each other. Marxist-Leninist ideology is therefore a synthesis rather than simply a sum of its constituent parts. But although relatively independent each of the components acquires its independence as an integral part of a whole. This wholeness finds multilateral validation and expression.

The wholeness of the components of Marxism-Leninism principally stems from the philosophy of dialectical materialism which has generalized the most significant achievements of the prior theoretical thought and has evolved into a science. It therefore serves to integrate all the other components of scientific ideology into a single unified system.

A more specific theoretical basis for this wholeness is derived from the (dialectically interpreted) principle of systematism, represented by concepts such as the "system quality" of a complex object and is aimed at understanding the multidimensional nature of these objects, the specific historical environment, the involvement of the evolutionary principle, etc.

The practical basis for Marxist integrity is its objective of transforming reality for the purpose of promoting free and comprehensive individual development of every member of society. The variety of life phenomena and processes represented by the philosophy of science leads to differentiation enhancing specific features of the components of Marxism and on the other hand, retention of their interrelations and the unfolding of processes to integrate these elements into a comprehensive theory.

Being inherently revolutionary, Marxist ideology is constantly developing. It is not a finite and closed entity but rather an open system receptive to change and consciously modifying itself to match the changing reality and the development of cognitive methods.

Third, defined in Marxism and verified by over a century of experience are certain theoretical principles vital to the development of conceptual components of a global development modelling system:

> contradictions as an endogenic source of development of the objective world in general and of social life in particular (contradictions between the level of development of the productive forces and the kind of production relations, as well as the struggle between opposing classes and between confronting socio-economic systems are primary sources of social development);

> the progressive nature of social development in which quantitative changes are combined with radical qualitative, revolutionary changes in

what exists at present (this progress may deviate from a strictly ascending line and involve intricate swings of history; however, this advance ultimately signifies a progressive approach to increasingly higher forms of social existence; this permanent upward motion is the basic source of the historical optimism inherent in Marxism);

the growing role of the human factor in history; the growing share of population actively and consciously taking part in historical events and processes; the increasing coherence of these processes, increasingly planned development counteracting sporadic factors; and

the imminent and highest priority of history turns out to be the comprehensive and coherent development of the human being, of all his creative faculties "without any other requisites but prior historical development that has been making an end in itself out of this comprehensiveness of development, i.e. the development of all human powers as such irrespective of *any present scale*" ([1], Vol. 46, p. 476). This list of arguments though incomplete, is sufficient to show that the theory and methodology of Marxism-Leninism complies with the requirements demanded of the basic concept of a global development MS. This conclusion is further validated when we consider the main components of the MS and the set of global-development alternatives and problems.

APPROACHING THE CHOICE OF BASIC COMPONENTS OF OBJECT OF MODELLING

The construction of an MS assumes the establishment of a general idea about the MS object as a real system. Since we are concerned with an MS of global development, the principal difficulty is that there are many possible principles for the selection of an MS object from the empirical reality. They are often simply taken as a list of global problems. A set of principal variables is then chosen to treat these problems. As they are interrelated, these principal variables can be described as a system. This approach (see [14]) has its disadvantages, as it is empirical and cannot ensure that the set of primary variables is complete. Both the list of problems involved and the set of parameters underlying the MS must be theoretically substantiated. I believe that it is the dialectical-materialist concept of activity which can serve as an initial theoretical basis.

No global problem can be treated beyond the context of human activity or as something independent of human beings or communities of human beings. In fact, any global problem is planted by and exists as a result of the activity of men and women united in a community. It is a totally different question whether or not they are aware of either the nature or the (possibly global) effects of their activity. Even if people adequately realize the nature of their activity, seldom can they do whatever they please, since the activity of every new generation is largely determined by the result of the activities of prior generations.

The choice of the dialectical materialist concept of activity is further justified by the fact that its origin and evolution cannot be divorced from the genesis and development of Marxism as a comprehensive theory. In the *German Ideology* the founders of Marxism wrote: "The premises from which we begin are not arbitrary ones, not dogmas, but real premises from which abstraction can be only made in the imagination. They are the real individuals, their activity and the material conditions of their life, both those which they find already existing and those produced by their activity" ([1], Vol. 3, p. 18).

In designing the object of a global development MS we can accordingly depart from three components: human beings, their activity, and the (natural as well as man-made) material conditions of this activity, the activity concept fundamental to the materialistic understanding of history being of utmost significance.

There are numerous definitions of activity in Marxist literature. One of the best is the definition which understands activity to be an inherently human form of active attitude to the environment, whose subject matter is changing and appropriately transforming the environment on the basis of acquisition and development of existing forms of culture [19]. It is generally known that growing (individual and collective) human needs are the source of human activity, that labour and, primarily, production are the main forms of activity, and that the ways, techniques, and means of activity are the central link in the structure of activity.

Yudin has demonstrated that the concept of activity belongs to the class of universal abstractions that combine empirical authenticity, theoretical insight, and methodological constructiveness. These concepts are few and "appear to consolidate the mental space of their time, provide this space with a motion vector, and determine to a great extent the type and nature of the subjects of thought generated by their time" ([24], p. 272).

This concept of activity well matches the nature of problems involved in global modelling. However, the latter is a specific domain in which all major functions of this concept are jointly implemented. Assuming that five or more different functions, each relating to an object structure are inherent to the concept of activity [24], the conclusion is that all these functions are implemented in constructing a global development MS. In this context, activity acts as an explanatory principle, an object of unbiased scientific study, a controlled plant, a design object, and an asset. This fact represents the comprehensive nature of global modelling itself, a trend in systems research.

Of course, this multifunctional feature is also inherent in other general concepts used in designing a global development MS. To avoid the inadvertent confusion of meanings, the creator of the model must therefore clearly define the specific function in which a given concept is used. Yet, this does not rule out that the same concept might be used at the same time in two or more functions. "Functional syndromes" of this kind encountered when applying a versatile concept in systems research are to be studied separately.

Now let us attempt to specify the activity concept as applied to the problem of extracting the main components of a global-development MS object. It should be pointed out initially that ways and means of activity are central to the activity structure. They relate the goal and the result, the ideal and the real. The quality of ways and means of activity determines the measure of agreement or disagreement between the goal and the result [21]. Ways and means represent the most dynamic component determining the direction and rate of the activity structure. Of these remaining components, two types, specifically: (a) subjècts and objects of activity and (b) relations between subjects are of fundamental importance.

Three principal types of subject activity are usually identified. They are individuals (human beings), social communities (historically specific societies and classes), and humanity. One more type has recently been added: cultural-and-areal communities (economic-and-cultural types, ethnoses and historic-ethnographic regions) [3]. The entire set of these activity subjects is covered by the concept of "civilization" as a historically specific socio-cultural entity [3, 16].

The primary object of human activity is nature as the unity of its three constituent spheres, i.e. the geosphere, the atmosphere and the biosphere. The human impact on nature has become so immense in the twentieth century that Vernadsky was able to speak of the biosphere turning into the neosphere [4]. Nature is an active rather than a passive object which exerts a feedback influence on the subjects in action. In fact, subjects are at the same time (though in a different respect) the objects of their own activity.

In light of what has been said above, it is reasonable to select the following types and forms of subject/object wholeness as a primary set of components constituting the object of a global development MS: nature (the geosphere - the atmosphere - the biosphere) - civilization (humanity - social communities - cultural-areal communities - the human being).

A different set of components is comprised of social relations and institutions that appear and evolve in the course of joint human activity. Of these, two main types can be reasonably distinguished in terms of global modelling problems, i.e. social relations and cultural social relations (in the broad sense of the words). Social relations and institutions involve economic relations, social class relations, political relations, etc., while culture has components such as material culture, language, art and science.

The two sets of components of a global development MS object are interrelated and can be represented by a matrix.

To summarize, the object of a global development MS can be described on the basis of the Marxist-Leninist concept of activity as a global system that (1) is a set of interrelated components of nature and civilization; (2) was generated by and evolves because of the activity of individuals, of social and cultural communities, and of all humanity; and (3) is aimed at satisfying their continuously increasing requirements. The material basis for the self-supported evolution of this system is provided by the progressive changes in ways and means of activity. A distinctive feature of this system is that the subjects of activity are numerous, and their goals are substantially different. This feature opens way to many alternatives of evolution for the system and/or its subsystems as well as to discrepancies between the subject's goal and the actual result of the subject's activity. Using an MS to study these alternatives is therefore of paramount importance.

THE SET OF PROBLEMS AND ALTERNATIVES OF GLOBAL DEVELOPMENT

Global problems are often dealt with outside any context, so that the impression is created that they form a complete list of problems that arise in the course of study of global system development. The actual situation is far more complicated. A problem is global if it is of concern to all humanity and demands concerted international efforts on a world-wide basis to solve it. These problems fall into two major groups, i.e. (1) transforming international relations and (2) optimizing interaction between humanity and nature.

The first group involves two sets of problems. One represents the contradictions of international relations in the military and political domain, such as eliminating the threat of a nuclear war, détente and reductions of arms and armed forces. These are the most urgent problems of today and the aspiration of resolving them has been uniting ever growing numbers of people around the world. The second set represents the contradictions of international relations in the economic domain, i.e. overcoming the economic backwardness of the developing countries of Asia, Africa and Latin America, and the struggle of the socialist and liberated countries to establish an international economic order based on the sovereignty and equality of all countries, justice and mutual benefit for all. This struggle is very closely linked to the struggle for peace and détente.

The second major group of global problems involves the dangerous imbalance of man-nature interaction and the necessity of assuring rational and planned utilization of the environment in which individuals and entire humanity live and function. In this group, the critical problems today are world population growth, food for this growing population, preventing especially dangerous diseases and negative effects of scientific and technological progress, meeting the world economy's mounting demand for energy and natural resoures, and protecting the environment against human-produced destructive effects.

Today, there are few who would dispute that the greatest difficulties in solving these global problems lie in the socio-economic and political domains rather than the scientific and technological. Even given that adequate scientific and technological solutions are available, any one of these problems is a challenging one, so that it can effectively be solved only by cooperation and the planned utilization of the resources of many countries, as well as by the consolidated efforts of all humanity.

Unfortunately, only some problems of humanity's evolution can be solved in that way. It is well known that the principal contradiction of modern times is the confrontation between the two fundamentally different socio-economic systems, capitalism and socialism. Nowadays their opposition is deep enough to permeate all the principal types of relations and structural wholenesses within a global system. No reconciliation is possible and any talk by bourgeois theoreticians about what they call the "convergence" of the two systems are just as groundless as ever (for further detail refer to [12]). (It is a different matter that the struggle between the two social systems can and must unfold in an environment of peaceful coexistence.)

Thus the present epoch has its principal contradictions of the evolution of humanity whose nature and solutions essentially lie outside the domain of global problems. This does not rule out their interrelation. In fact, they are closely related, which demands that any global problem be treated not as an end in itself, but rather in the context of a wider system.

The above description of a global system and its constituent components enables the domain of global-system development problems to be represented as a multilevel structure with both feedforward and feedback couplings between levels (Fig. 3).

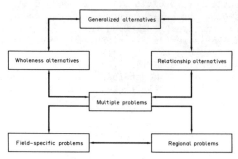

Fig. 3. Problem levels of global system.

The first, i.e. top level, is formed by generalized development alternatives of the global system as a whole, i.e. the global system as an integrated set of its principal components. The second level includes two types of alternatives, one designed in accordance with the typology of the subject/object wholenesses, and the other in accordance with the typology of global system relationships and institutions. The third level encompasses complex problems generated at the intersections of wholeness alternatives and relationship alternatives. Finally, the fourth level represents field-specialized problems (i.e. those corresponding to the principal branches of human activity) and regional problems (in which the problem setting of the previous levels is modified by specificity of the region).

The presence of both feedback and feedforward interlevel couplings makes these levels a strongly coupled system. Each type of problem in this system is to be treated in close connection with the entire set of global system problems, and as a manifestation of "emerging properties"[2] of the global system.

This interrelationship requires that specific problems and alternatives representing the various subsystems of a global system at the current stage of development be taken into consideration alongside the general properties of the global system if a generalized alternative is to be constructed. On the other hand, specific problems and alternatives can only be chosen appropriately by proper consideration of the generalized alternatives and as a specialization thereof. This fact reflects the inherently contradictory and paradoxical nature of systems-oriented thinking [20].

Incidentally, there are prerequisites for circumventing this paradox dialectically. On the one hand, Marxist-Leninist theory accepted as the fundamental concept for a global development MS provides the basic principles to create a general idea of what the structure and development trends of this global system might be. On the other hand, the scope of specific development problems can be derived from empirical data. One is then left with the task of adequately correlating the specific problems and the general concept to eventually construct a set of problems and alternatives for contemporary global development.

REFERENCES

1. K. Marx and F. Engels, *Collected Works*, 2nd edn, Vols. 20 and 46, Part 1, in Russian.
2. I. V. Blauberg, Wholeness and systemicity, in: *Systems Research Yearbook 1977*. Moscow, Nauka (1977), in Russian.
3. Yu. V. Bromley, On the relationship between cultural-areal communities and civilization, in: *Sociology and Problems of Social Development*. Moscow, Nauka (1978), in Russian.
4. V. I. Vernadsky, *Naturalist's Reflections. Book two: The Scientific Thought as a Global Phenomenon*. Moscow, Nauka (1977), in Russian.
5. D. M. Gvishiani, Methodological problems of global development, *Voprosy Filosofii* No. 2 (1977), in Russian with English summary.
6. D. M. Gvishiani, Global modelling: complex analysis of world development, *World Marxist Review* No. 8 (1978).
7. V. A. Gelovani, A global development modelling system, in: *The Methodology of Systems Analysis*. Moscow, VNIISI (All-Union Institute for Systems Research) (1978), in Russian.
8. S. V. Dubovsky, A system of global development models, in: *Methodology of Systems Analysis*. Moscow, VNIISI (All-Union Institute for Systems Research) (1978), in Russian.
9. S. V. Emel'yanov and É. L. Nappel'baum, Basic principles of systems analysis, in: *Problems of Management in Socialist Industry*. Moscow, Ékonomika (1974), in Russian.
10. V. V. Zagladin and I. T. Frolov, Contemporary global problems: Sociopolitical and philosophical aspects, *Kommunist* No. 16 (1976), in Russian.
11. V. V. Zagladin and I. T. Frolov, Contemporary world problems as viewed by communists, *Problemy Mira i Sotsializma* No. 3 (1978), in Russian; a parallel English edition, *The World Marxist Review*, is also available.
12. V. V. Zagladin and I. T. Frolov, Global problems and the future of humanity, *Kommunist* No. 7 (1979), in Russian.
13. É. M. Korzheva, Simulation of culture: methodological approach, in: *Methodology of Systems Analysis*. Moscow, VNIISI (1978), in Russian.
14. N. I. Lapin, Problem of social indicators in a model of a global study, in: *Systems Research Yearbook 1978*. Moscow, Nauka (1978), in Russian.
15. N. I. Lapin, Topology of non-formalized components of a modeling system, in: *Methodology of Systems Analysis*. Moscow, VNIISI (1978), in Russian.
16. M. Mchedlov, Evolution of communist civilization, *Kommunist* No. 14 (1976), in Russian.
17. N. F. Naumova, On the problem of goals in systems-based global models, in: *Systems Research Yearbook 1978*. Moscow, Nauka (1978), in Russian.
18. I. B. Novik, *Simulating a Complex System*. Moscow, Nauka (1968), in Russian.
19. A. P. Ogurtsov and É. G. Yudin, Activity, in: *Great Soviet Encyclopedia*, Vol. 8. Moscow (1970), in Russian; English edition published by Macmillan is available.
20. V. N. Sadovskii, *Foundations of General Systems Theory*. Moscow, Nauka (1974), in Russian.
21. N. N. Trubnikov, *Concepts of Goal, Means, and Result*. Moscow (1968), in Russian.
22. P. N. Fedoseev, *Philosophy as a Component of Ideology*. Moscow, INION AN SSSR (1978), in Russian.
23. S. Ya. Shcherbov, Accounting for age-and-sex structure of population in a global model, in: *Computer Programming. Problems of Socioeconomic Systems. Models in Operations Research*. Moscow, VNIISI (1978), in Russian.
24. É. G. Yudin, *Systems-Theoretical Approach and Principle of Activity*. Moscow, Nauka (1978), in Russian.

25. G. A. Albizuri, What future (if any) may global modeling have?, in: *Proc. Seminar on Large Scale System and Global Models*. Dubrovnik, Yugoslavia (1978).
26. S. Cole, *Global Models: An Evaluation of Their Relevance to Policy*. University of Sussex, U.K. and UNITAR (1978).
27. W. Leontieff et al., *The Futures of the World Economy*. United Nations (1976).
28. M. Mesarovich and E. Pestel, *Mankind at the Turning Point*. New York (1974).
29. M. Ward and H. Guetzkow, *Challenges to Integrated Global Modeling: Paper for 6th IIASA Global Modeling Conference*. Luxembourg (1968).
30. *World Futures: The Great Debate* (Ed. C. Freeman and M. Jahoda). London (1978).

A MODEL OF THE REPRODUCTIVE SYSTEM: CONCEPTUAL CATEGORIES

Yu. A. Levada

A major feature of a sociocultural system is its ability to reproduce in the course of time its organization (or structure in the broad sense of the term). This process involves a continuous change of the system's human and "material" components whose characteristics including those of the relevant environment may vary infinitely. Such self-reproduction may be viewed as the most general result of the system. "For bourgeois society treated as an entity the ultimate result of the social reproduction process is always society itself that is man in his social relationships. The process of production as such operates only as an aspect of time" [1, p. 222]. This holistic entity is presented here as a reproductive system (RS) in a model which employs the categories of memory and time in the time-oriented mechanism of system organization. The purpose of this chapter is to discuss these categories. In the discussion that follows primary emphasis will be placed on the analysis of the interpretation, application and interaction of the categories listed above.

The temporal organization of a system implies the existence of special mechanisms responsible for the recording, storage, transformation and implementation of characteristics associated with past states of the system (this is what distinguishes time-oriented organization from "continuance" in time, specific to inorganic systems). In the context of the RS model that we are concerned with such functions can be identified at the two extreme levels of system organization - societal and personal. Only these levels possess a specialized "long-term" memory mechanism which maintains the continuity of the structure in time. (In the history of social science a structure of this kind has been traditionally viewed as "drawn" to either pole which defined the methodological approaches of sociologism and social psychologism, respectively.) At the societal level the memory functions are institutionalized in superorganism-level cultural structures while at the personal level they are materialized in psychophysiological memory mechanisms. The complex interaction of these extreme levels of organization makes the personality of an individual in society the most stable sociocultural institution. For this reason models of social and individual memory mechanisms may be viewed as similar only within explicitly defined limits.

The state of a system which possesses a memory mechanism (regardless of what kind of mechanism it is) cannot be fully determined at a given point in time by the immediately preceding state (i.e. the system is not Markovian). The memory mechanism is associated with a more complex type of determinism where the present state of the system is conditioned by some aspects of *different* past states as well as by their projections into the future. It follows therefore that a temporal system should have at least two time scales: for measuring the time sequence of states (continuance) and for time structures recorded by the memory.

The operation and reproduction mechanism of the RS model may be presented as a model which includes programmes of two types: (1) reproduction of a given pattern of its organization in time and (2) achievement of a specific effect when exposed to many alternative courses of action (in "strategy selection space"). This differentiation is of fundamental importance as it is purely analytical.

The first type of programme is aimed at maintaining a specific pattern which records the most essential aspects of the structure of society or an individual. Its function is to eliminate such values of these variables which go beyond the acceptable limits (these latter may be defined in different ways and/or degrees of rigour). In particular the correspondence of societal and personal characteristics may imply only the existence of a more or less strictly defined range of possible values of such structures, i.e. the values of their "disparities". The pattern maintenance mechanism fixes a certain reference (typically, an ideal past state) and the current values of the model variables are compared with the reference pattern. In this sense the programme is looking into the "past".

The second type of programme, the strategy selection (optimization) programme, is built around a specific figure of merit that will either enable the system to reach the required level of performance or maximize this parameter or produce the maximum effect at minimum cost in terms of the resources consumed (the first of these variants may be called "logical", the second "technological" and the last "economic"). The optimization programme must provide for a range of means (strategies) defined in some space which when implemented achieve a certain goal (the state defined by the selected criterion of optimality). A programme of this type is oriented towards a possible incompletely defined future. Finally the optimization programme is capable of self-improvement and can select increasingly effective strategies (a self-learning system).

Several researchers who have applied systems analysis methods to biological systems have concluded that models must take account of the dual organization of such systems. Two types of memory - "permanent" and "fast" - are usually distinguished. According to Geodakyan a memory of the first type serves to maintain "the perfection of biological organization": at various biosystem levels these functions are supported by DNA, the cell nucleus and by female organisms. The fast memory is responsible for interaction with the environment which produces progressive changes. These functions are performed by protein, cytoplasm and by male organisms [7, p. 375] [8].

In sociocultural systems functions that to a certain approximation may be considered identical are implemented by mechanisms of a different nature. In the context of the RS model under discussion the programme which maintains the structure in time may be classified as a *programme of culture* while the optimization programme may be viewed as that of *experience*. Each is associated with a specific type of memory. In the first instance this is a long-term, accumulated, generalized memory which sets the basic framework of the system's existence. It contains normative standards, values and estimates of the performance of RS agents. The other type of memory is a fast "live" and hence mobile mechanism whose contents include conditions and means to perform specific goal-oriented actions (in the final analysis the criteria of optimization are associated with the cultural programme and hence make up part of the long-term memory).

By definition the two programmes are interrelated: the normative values of the programme of culture determine the objective functions of the programme of experience, the operational mobility of strategies (means) maintains the stability of the goals and so on. Transition from one programme type to the other is less straightforward. If the programmes are distinguished along the most general analytic lines, the model cannot cover the process of their transformation. The programmes of interest to the analyst should first be elaborated and sectionalized, if such procedures as the "conversion" of normative values into tools or conversely of tools into an end in itself are to be examined.

The concept of the "cultural programme" viewed as an integral entity cannot be applied to represent such processes to which we will refer as normative transformations. Specifically it would be necessary to evolve adequate forms for the model representation of intracultural dichotomies of central or peripheral structures including subcultural and countercultural, institutional and personal, dynamic and stable, etc.

Within the scope of the definition of culture, far from rigorous but generally accepted in sociological and anthropological literature (see [5]) the programme types classified above are similar to the descriptions of the normative and instrumental levels of the systems of culture. Theoretically and also historically another, "authorizing" level may be distinguished which is explicitly reflected in mythological consciousness. Without going into the specifics associated with the cultural values of each level thus defined it would be proper to emphasize that the normative and value structures dominate the model of culture.

The well-known concepts of "two cultures", technical and humanitarian [9] "superculture" and "tradition" [13] and similar dichotomies can be described to a sufficiently high degree of approximation using the conceptual formalism discussed in this chapter as in effect we are dealing here with the correlation between the programme of culture and the programme of experience. The traditional "humanitarian" approach tends to treat instrumental experiments in terms of stable ("eternal") normative categories of culture. The technical approach tends to re-evaluate and reduce them to "technical" means for the solution of social and cultural problems. The tradition to oppose one approach to the other apparently dates back to Plato who drew a distinction between "wisdom" and "artifice" ("techne") [6, Vol. 3, Part II, pp. 160; 484].

In the context of the conceptual model under discussion it would be appropriate to examine separately changes in the instrumental area of "experience" and in the normative "nucleus" of culture. A reconstruction or re-evaluation of the cultural paradigm (the system of cultural values) cannot be considered to be an immediate result of changes in instrumental-technical structures (by analogy with biosystems which do not inherit acquired properties). As the reproductive system functions acts of "inheritance" occur in each programme independently. Thus the instrumentalization of elements of the "nucleus" structure of norms and values cannot be brought about by the development of "technology" as such (in the broadest sense of the word), since it is a product of nuclear processes proper. In other words this process is stimulated not by the growth of "technology" but rather by the changing role of the instrumental programs in human life.

Although in the above discussion RS programmes were defined in terms of their memory mechanisms they should not be viewed as information storages or data "containers". An effective RS model must account for such memory attributes as selectivity (as regards both recording and retrieval of relevant information) and activity (the normative meaning of information). Information "storage" *per se* in the RS long-term or fast memory implies its transformation (rearrangement, re-evaluation) stimulated by the reciprocal influence of memory layers and structures, by the projection of current experience, etc.

Presentation of the RS memory at the different levels of consciousness constitutes a specific problem. Rewording Teilhard de Chardin's well-known maxim "an animal knows while man knows that he knows" (10, p. 165) one could say that man as a being that belongs to a genus remembers that he remembers some significant part of his culture and experience. True enough he does not remember it all in a similar form or equally well. Human consciousness multiform as its structures and levels performs an important function in presenting to man or society his or their "own memory". RS types differ in particular in the scope and mechanisms of this procedure.

Not only the reproductive system and its memory but also the realization of this memory is time-structured, extended in time and relates events to specific instances which are fixed for this purpose. It is precisely this factor that accounts for the significance and polyvalence of the category of time in RS models (we are concerned here not with philosophical analysis of time or psychological examination of its perception by man but exclusively with the possibility of modelling time characteristics which "operate" in the RS structure.

Historical and cultural studies of this problem (3, 14, 15) suggest the existence of opposition between "natural" and "artificial" (mechanical) measures of time. The former also include the temporal characteristics of the life cycle (generations, periods, etc.) and parameters of external space rhythms (year, season, month, day) internalized in one way or another. The time parameters of mythological events constitute some kind of supernatural opposition to natural measures of time. In the context of our model approach it becomes possible to identify within the RS framework "functional" "mythological" and "technical" meanings of temporal parameters. The "functional" time scale comprises measures of the active life of sociocultural systems and of group or individual social activity. The time scale in this instance is defined by the limits of conscious memory and attention. Quantitatively they are associated with the range that extends from "generation" rhythms [12, p. 6ff] to the rhythm of an individual action. The "mythological scale" is in a sense the projection of functional time parameters on to the plane of the corresponding type of consciousness. It legitimizes the time frames of events which are taboo in "normal" conditions such as life spans, periods, etc. (cf. ancient Hindu Vedic concepts of world cycle mega measures which exceed by many orders of magnitude not only "conventional" time parameters but also any astronomic chronological scales known to modern science [14, p. 80]. "Technical" time scales relate purpose-designed or specially introduced systems (mechanical, subatomic, etc.) which exhibit different periodicity. Dimensionwise all the scales discussed above may of course overlap. However, functional time may apparently be based only

associated with conscious human action (like the "measure of man" of the
Hellenes). In relation to these both the ancient Hindu mega measures and
analytic "micro measures" (their interpretation in the Middle Ages is
discussed in [2, p. 72] are but secondary.

Any time scale is built around the concept of a rhythmic structure or cyclic
process which may be regarded as an elementary "unit" of time organization.
In historically known world models it may be interpreted as an ultimate
"unit" and thus related to the whole organized universe (in the primary
meaning of this category). Any linear sequence of states is not only
measured in terms of cyclic units but may also be treated as a singular
cycle with an infinite or an indefinitely long period. Empirical analysis
of different processes reveals a combination of cyclic and vector changes
(in a manner of speaking, a "wheel" rolling along a "vector" path). By way
of example we may mention production cycles in economic development or
biological, psychological or social rhythms against the background of
linearly oriented processes of energy and genetic entropy. It should be
borne in mind that any philosophic or epistemological discussion of the
relationship between cyclicity and linearity in historiosophy and methodology
of history is beyond the scope of this article. Here we shall address
ourselves only to problems associated with the model representation of some
aspects of time structures.

Temporal organization of a goal-oriented action as an operation is typically
presented as a linear sequence of states which exhibits a certain trend.
The time count starts from a finite point and the model is therefore
organized as a finalistic entity. It is symmetric to the model of causal
deterministic interpretation known in historical science where the time
count begins from some "initial" state, from "zero". The model is
essentially quasifinalistic. The functions and prestige of a rational
goal-oriented action or operation in social life clearly account for the
emphasis on linear models so common in modern historiosophy. Extrapolation
of a rational goal-oriented action model to a macrohistorical scale can be
found for example in Hegel's philosophic models of secularized theodicy and
in some concepts of positivism.

In view of the fact, however, that a rational action invariably involves the
execution of a specific programme it cannot be adequately described by means
of a linear model. To record such a programme ("memory") would require
a different time scale (a different time sequence "line") and an appropriate
transition mechanism, i.e. another and differently directed "line" of
movement in time. A linear time model is just as inadequate for presenting
causal determinism. Thus, historians have established that changes in
different areas of social life in principle do not occur at the same points
in time (see, for example, [11, p. 127] Brodel's views on the correlation
of "fast" and "slow" history, although the latter is seen by the author
exclusively as the history of economic development).

The multiplicity of time scales is a mandatory condition for the existence
of a reproductive system which reproduces its pattern of organization in
time and thus programmes its subsequent states (obviously when applied to an
RS which represents empiric reality, "programming" may be understood only in
its most general and probabilistic sense as the setting of a possible range
of favourable conditions or acceptable changes). This conclusion applies
to an idealized RS, i.e. in this situation we do not take into account the
system's cultural and social heterogeneity. Such parameters should they be
included in the model would "multiply" the lines of temporal organization

still further. The relation between social structure heterogeneity and the multiplicity of "social times" was examined in depth by Gurvitch [16]. Here it is essential to emphasize that social time "multilinearity" (and "multicyclicity") is inherent even in the homogeneous model of a sociocultural system. As we deal here with a time-organized system, its "eigentime" is a system of times.

The "space" concepts of time which appear to accommodate different coexisting time lines and time frames have attracted the attention of scholars concerned with archaic social patterns [3, p. 90]. It may be concluded from the above that the space model of a system of times is to be found also in more advanced sociocultural structures (of course the "spaces of times" we have in mind here are dissimilar).

It should be emphasized that the preceding discussion does not apply to physical time which through its irreversibility links up the material substrates of any social cultural or psychological processes. The time characteristics treated here may be related only to the "phase" time of different RS structures and to subjective or more precisely "socio-cultural" time. These are the time parameters perceived by social consciousness and reflected in cultural values. It is in this context only that one may legitimately speak about nonhomogeneity, multivalence or multidirectivity of time scales and measures. Thus events that are meaningful to a specific cultural structure or social individual cannot be treated within the framework of a uniform-scale model: given equal external, say, astronomical or other measures, some events appear to be closer than others when perceived "socioculturally", and some time spans may well look longer than others. In the sociocultural time system events and time intervals are evaluated in terms of their relationship to special "marked" points which may be called "axial" if the connotation of the term introduced by Jaspers is extended somewhat. In historiosophic models such properties, i.e. symmetrically ordered time characteristics are assigned to certain "turning" epochs or events [17] while in rational goal-oriented programmes it is the "ultimate points" that are pivotal. A broader meaning is associated with the axis of contemporaneity which is inherent in a culture and defines the time orientations of the system with reference to the "contemporary" situation that has events of the meaningful past and of a probable future projected on it. In Jaspers's words events are evaluated in terms of their relationship to what is significant "here and now".

RS models possess specific memory update mechanisms or procedures for presenting a transformed past to the present, to the current sociocultural system. Thus, in the most archaic social structures the past recorded in the structure of epochs-heroic, mythological, or some other or in the personalities of immediate ancestors and represented by appropriate human characters coexists with modern day-by-day life [3, 14, p. 112]. However, this situation can hardly be viewed as "identity of times" since the most archaic cultural layers exhibit a more or less distinctly visible boundary between the updated memory and the present reality. The ghost of Hamlet's father who archetypically belongs to such layers indeed exists in the present but only as a ghost.

In social structures of higher complexity culturally identifiable periods, events or individuals of the past are to be found in the present as normative symbols (instructive, inspiring, etc.). We may visualize a situation where this symbolic meaning has been lost and the linkage of "times" (generations, epochs) is confined to the preservation and transfer of practical experience such as is scientifically organized and recorded in the practical experience such as is scientifically organized and recorded in appropriate objectivized texts. Under this hypothesis which apparently has no parallel in real RS prototypes the program of culture is reduced to the programme of experience with associated transformation of the system of time scales.

Only by representing the time organization of a reproductive system as a complex multilevel hierarchy would it be possible to describe conceptually the structure capable of recording the states passed by the system as past ones (referenced to appropriate time frame axes) although still significant at present. This structure readily provides multiple access to the cultural content of the past, its interpretation and re-evaluation and hence its continuous updating. The limits of this process define in effect the boundaries of a time structure which may be classified as a *historical* system.

REFERENCES

1. K. Marx and F. Engels, *Collected Works*, Vol. 46, Part II, in Russian.
2. J. S. Bruner, *Beyond the Information Given. Studies in the Psychology of Knowing*.
3. A. I. Gurevich, *Categories of Medieval Culture*. Moscow, Iskusstvo (1972), in Russian.
4. M. Koul and S. Scribner, *Culture and Thought. A Psychological Introduction*.
5. E. S. Markaryan, *On the Genesis of Human Activity*. Erevan, Academy Press (1973), in Russian.
6. *Plato Laws, Collected Works*, Vol. 3, Part II. Moscow, Mysl' (1972), in Russian.
7. *Evolution of the Concept of Structural Levels in Biology*. Moscow, Nauka (1972), in Russian.
8. *Systems Research Yearbook 1976*. Moscow, Nauka (1976), in Russian.
9. C. P. Snow, *Public Affairs*. Moscow, Progress (1972).
10. Teilhard de Shardin, *Le phènomene humain*. Paris (1959).
11. *Philosophy and Methodology of History*. Moscow, Progress (1977), in Russian.
12. E. V. Stahl, *Odyssey - A Heroic Poem of Wanderings*. Moscow, Nauka (1978), in Russian.
13. K. E. Boulding, *The Impact of the Social Sciences*. New Brunswick, Rutgers (1969).
14. *Cultures and Time*. Paris, UNESCO Press (1966).
15. *The Future of Time*. New York, Doubleday (1971).
16. C. Gurvitch, *Les cadres sociaux de la connaissance*. Paris, PUF (1966).
17. K. Jaspers, *Vom Ursprung und Ziel der Geschichte*. Zürich, Artemis (1949).

ON THE PROBLEM OF SIMULATION OF SYSTEMS WITH CHANGING STRUCTURE

V. I. Danilov-Danilyan, I. L. Tolmachev and V. V. Shurshalov

TWO APPROACHES TO OBJECT DESCRIPTION

A great number of different methods are used in the theory and practice of socio-economic modelling. With certain reservations they can be divided into two classes: those based on the indirect and those based on the direct description of the object to be modelled. In the first category we can place programming, multisectorial input-output models, production functions, various correlation and regression techniques, etc. In the final analysis all are intended to describe a reality in terms of a set of mathematical formulae, however complicated and diversified the problems to be solved to derive them might be. The language of this indirect description has only four basic semantically different components: the unknown, the given (or parameter), mathematical operation and constraints (for instance, equalities, inequalities, implicative relations). The model in this case is not structurally similar to the simulated process, although in some aspects (with respect to model modularity, for example) this correspondence is possible.

To illustrate this, we can consider programming problems involved in modelling material production. Variables of these problems are interpreted as production capacity utilization levels of different production processes. To construct a model of this type, additional effort is needed to partition the process into its artificially isolated components. It is with these components and not with the process as a whole that the model variables are made to correspond. If the process is branching it is described not by one but by many variables, with some of them referring to adjacent production processes simultaneously. Hence there is no unique correspondence between the set of model variables (or any of its subsets or quotient sets) and the ensemble of production processes or operations involved in the simulated activity. This indicates that there is no similarity or homomorphism in the structures of the model and the reality, and it is precisely because of this fact that descriptions of this class are referred to as indirect.

The models of the second class are based on the direct description of the simulated process. They are distinguished by their homomorphic representation of the structure of the simulated reality. They are used in computer simulation, in business (simulational, situational) games, and sometimes for physical modelling (e.g. using analogue computers). When direct description is employed, the object to be modelled is considered as a system (the simplest and most widely used definition of this notion is sufficient: a system is a set of interacting components). The system structure may be presented as a graph whose vertices correspond to system components, whereas the arcs correspond to the potential links between them. Both the vertices and arcs of the graph may be characterized quantitatively (by their capacities, throughputs, etc.). To describe a system, it is certainly not sufficient to merely specify its structure. The components may also be regarded as systems of another level of aggregation. The choice of the latter is determined by many factors, some of which will be briefly dealt with subsequently.

Let us assume that the aggregation level is fixed and the process description, i.e. description of its structure by a graph, is available. The processes to be simulated may be of widely differing nature: production systems, decision-making systems, planning processes, social processes, etc. The material implementation of the structure of these processes ensures that part of the necessary conditions for its functioning will be fulfilled. On the other hand, the model structure establishes the ranges of variations for the characteristic parameters of its functioning. An approach of this nature to process description makes different models of the same process structurally similar by definition.

In some trivial cases, the process description - indirect by intention and the means used - may turn out to be isomorphic to the direct description, and have a structure similar to that of the simulated process. This is often the case with models of physical and mechanical processes. But we will deal with conceptually different cases where the simulated process is so complicated that the indirect description isomorphic to the direct is either wholly unattainable, or nonoperational (i.e. a model based on a description of this kind can neither be qualitatively analysed by the available mathematical means nor numerically solved by available computers).

We will not go further into detail about the complicated epistemological problems of the modelling theory, except for a few passing remarks. Assume that there are several verbal descriptions of process by a differing level of aggregation: $D_1(A)$, $D_2(A)$,..., $D_s(A)$ (lower indices correspond to lower levels). For each of them, there is a corresponding model based on direct description $P_1(A)$, $P_2(A)$,..., $P_s(A)$, so that $P_i(A)$ represents the structure of the process identical to $D_i(A)$, and for $i < j$ the structure of model $P_j(A)$ is homomorphic to the structure of $P_i(A)$. As for the model based on the indirect description, there may be no direct correspondence to any of the verbal descriptions available. To be more exact, if $Q(A)$ is such a model, a homomorphic representation H of model $Q(A)$, or rather, of its output, is sufficient, so that for some i $HQ(A)$ is structurally identical to $D_i(A)$.

The model based on indirect description is always a formal structure which can easily be analysed by mathematical means, whereas the model based on the direct description implies experimentation (either numerical, using a computer or of simulation gaming type involving a group of experts, or else, mixed, man-machine type). The construct based on the direct description is not yet a model by itself (in narrow, "classical" meaning of the term). It is a certain real system structurally similar to the modelled process within which the model "emerges" in the course of the experiment.

Models of both classes have certain advantages and disadvantages. The strength of indirect description techniques is its universality, its readiness to enlist the entire analytical arsenal of modern mathematics. However, in management practice the use of these methods is hindered by the very high demands they make on the user's expertise. In fact, it will be successful only if the user is as knowledgeable in the art of modelling as the model designers themselves. To a practising manager intimately versed in the behaviour of the system to be modelled, the language of the model appears to be highly restrictive and unnatural, and he has to adapt to it - which may prove ultimately unfeasible. The development of special automated interface mediating between the user and the model often involves complicated problems of programming, semantics and data retrieval, and is not always feasible economically even when theoretically the problems can be solved.

Models with direct description are far more convenient for the user, as their inner language practically coincides with the natural language of object description at a corresponding level of aggregation. This creates extensive possibilities of using human beings (with their inherent diverse and non-formalizable behaviour) as components of the model itself (in game simulation or man-machine simulation systems). These possibilities do not always exist for the models based on the formalized indirect descriptions.

An important class of direct description models arises from application of Monte Carlo (statistical experimentation) techniques. These methods are used for dealing with both deterministic (e.g. evaluation of definite integrals) and stochastic problems. In the latter instance, the solution yields some average characteristics of the process, and to make these estimates statistically reliable, a certain number of runs is needed. With a complex system the number of runs required to assert the statistical significance becomes prohibitive. In such instances models similar to Monte Carlo models are often used to determine not only the average parameter values, but rather its behaviour in extreme conditions, for instance, for peak loads, maximum possible deviations of its inner parameters, etc. With certain reservations, incompletely formalized simulation models may be considered "non-strict" (i.e. not satisfying statistical criteria) models of the Monte Carlo type.

Of all the models using the indirect object description, we will go on to detail only computer oriented simulation models. The fact of whether the model has been completely computerized or is of man-machine type is immaterial for our further considerations.

ANALYSIS OF STRUCTURAL CHANGES THROUGH SIMULATION

The isomorphism of structures of the simulated object (at a certain level of aggregation) and the direct description model implies that the latter is designed for handling problems of running management and short-term planning. It is in these problems that the structure of the process to be controlled is assumed as invariable. This (in accordance with the adopted presentation of a structure as a graph) implies that the range of system components does not alter, and no new link between them may emerge. Consequently, a simulation model can by itself be used to study the process as it is, but not in its evolution. But models based on the indirect description can be successfully employed in the analysis of the evolution processes as well. For instance, many economic-mathematical models using mathematical programming, production functions or input-output approach are used for analysis of investment processes, changes in the economic structure, etc.

Let us consider a complex system of a transportation network type. A
simulation model can be used to study the system functioning for a wide
range of internal and external conditions. However, when discussing the
further development of a transportation network (design of new terminals or
lines) new simulation models - one for each option considered - have to be
created. The alternatives can thus be studied, and various socio-economic
indices assessed supplying the information required for decision-making.
Creating the appropriate software package is a challenge. Modifying the
basic simulation model for the initial version of the transportation
network, i.e. updating the respective software, is far from simple. As a
matter of fact, each new terminal or line entails changes in the description
of many of the components of the currently existing system. The modifications
do not reduce to simply local changes, but involve a multitude of system
components and interconnections, and thus a wholesale reconsideration of the
entire program package is required.

Controlled process adaptation to the changing environment does not boil down
to mere alterations of its inner structure. It is possible to actively
influence the environment through the channels of communication between the
system and its environs. For instance, if there is a discord between the
supply of consumer goods and the demand for them, we can, by changing the
inner structure of demand, control the supply. Price control also has a
direct effect on the production system and on consumer demand. Processes of
adaptation by actively modifying the environment are studied using the
simulation model with invariable structure in accordance with the
conventional practice of using such constructs [2, 3]. However, a special set
of tools is required in order to efficiently analyse different options of
the inner structure changes.

Before specifying the problem, it should be noted that short-term planning
is generally considered as involving tasks concerning the utilization of
available production capacities, and actually (but not potentially) available
production resources. It is a question of system functioning within the
specified production structure. Long-term planning deals with the
development, expansion or reconstruction of capacities, and the actualization
of potentially available resources. In this case, the focus is on the
analysis and choice among the versions of structural changes.

Let us assume that there are a system with a fixed (so far) structure, its
description at a certain level of aggregation, and a simulation model with
the isomorphic structure. Possible changes in the structure of the
simulated system must be followed by isomorphic modifications in the
structure of the simulation model, and the problem is to develop a procedure
enabling us to perform modifications of this kind readily and at a low cost.
In other words, we have to create a "superstructure" over the simulation
model, thus transforming it into an even more complicated construct which
we henceforth will call *simulation system* (acknowledging the sharp
distinction between simulation system and *simulation models* in our
terminology).

A simulation system should provide the software needed for direct
experimenting with a range of simulation models, i.e. a basic model, along
with all its modifications. (Naturally, the attribute "basic" is somewhat
conventional. It may be justified only through reference of the isomorphism
of its structure to the structure of a real system as it emerges at the
moment of the experiment. However, when a system is to be studied at the
design stage, it may happen that no version under consideration can claim to
be singled out as basic.) To orient a simulation system software package

towards a considerably wide range of application, if not general purpose, first of all a unified language of simulation model representation has to be developed. This is the only thing that may serve as a genuine foundation for the definition of structure modification operations. Sufficient rigour in definitions of these operations is a prerequisite for the development of software that will be flexible enough and will have an extensive scope of potential applications.

The language of automata theory (or alternatively, of Turing machine theory) turns out to be a sufficiently universal instrument for the representation of complicated process simulation models [8, 7]. Each component of the simulated system (represented in the graph by a vertex) is treated as an automaton with a finite or infinite set of inner states and a finite number of inputs or outputs. From the computer science point of view, informational rather than computation problems are the principal hurdles in this approach. Incidentally, this is almost always the case in practice as well. Since simulation models are based on the direct description of the simulated process, they seldom give rise to complicated computational problems, unlike models based on indirect description. The latter always involve problems of optimal programming, systems of linear and nonlinear algebraic and differential equations (both ordinary or partial), etc. And every problem mentioned above in the case of high dimension often proves not simply difficult computationally, but totally unsurmountable. These considerations bring us to see in a true light the real significance of the informational problems in simulation, since we have to simulate not only the functioning of the simulated system components (as represented by automata) but their interactions (by piecing this automata one with the others) as well. To be in a position to cope with the situation, we have to be able to deal with the problems of classification, accumulation, storage, sorting and routing of information in a simulation model. The efficiency of the solution of these problems greatly influences the merit of the model, its speed, the number of experiments that can be performed, and reliability of the results. Hence the development of a high-quality information service software is a *sine qua non*, a proviso, in transforming simulation model into a simulation system.

It should be noted that simulation models vary considerably with respect to the information servicing they require depending on the nature of the simulated process. The simplest information problems are dealt with when simulating physical processes characterized by a fixed set of interdependent parameters. In this case it is parameters which are to be represented by graph vertices (automata) and the arcs of the graph characterize their interdependence. As for the automata, they are of the functional (in stochastic simulation, correlational) type. In such cases, not only is the structure of the simulated system and simulation model rigidly fixed, but the information pattern is as well. By specifying the initial state of a system, one can simulate its dynamics. And in the process, the pattern of information characterizing a system will not change. Even the dimensionalities of all data files are set along with the model structure once and for all. However, simulation modelling of social, economic and some technological processes is of a different nature. In this case, if the structure of the simulated system and the model itself remain fixed, the information pattern may vary. Information may accumulate and the nature of this accumulation may not be foreseen; it can be revealed only in the process of the model's operation. These are exactly the cases which present concrete difficulties for the development of information service software.

Simulation of Systems with Changing Structure 139

Changes in the structure of the original system are manifest in the apparition of new entities in it, in the elimination of some of the existing ones, or their replacement by other entities (the last instance can obviously be regarded as a combination of the first two). Accordingly, in a simulation model the changes are to be represented by the introduction of new, and by the removal or replacement of the currently available automata, i.e. vertices in the graph representation of the structure. Removal of an automaton is accompanied by the severing of all links connecting it with other automata or in graph-theoretical language: vertex elimination should be accompanied by the elimination of all incident arcs. An inverse problem arises with the introduction in the structure of a new entity: an automaton is to be added together with all its links, and a graph vertex along with its incident arcs. These are the procedures which the simulation system must perform over a simulation model automatically using standardized operations. The computation procedures involved are obviously very specific. The main problem lies in modifications of the information exchange, of the information pattern of the model. Below we will examine one of possible approaches to deal with the problem based on the standardized representation of the simulation model.

SIMULATION MODEL-BUILDING PRINCIPLES

Entities of the real world are interconnected via material, energy and/or information flows. In the case of direct descriptions all these connections are represented by some informational (an analogue models - energy) flows between the models of the entities. Thus, links of diversified nature are represented in models by the same means.

We will deal further with a system whose components ("entities") are interconnected and connected with an external world only by these informational links. Let us assume that the system structure is described by a graph (or rather multigraph) with the vertices representing its components, and the arcs describing the links between them. Let $U = u_1, \ldots, u_M$ be a set of vertices; if there can be a link between the components represented by u_i and u_j this is to be reflected by a properly oriented arc d. Any two model vertices may be interconnected by several arcs, which may happen not only when the corresponding "physical" components are connected by the links of a different nature. Sometimes it is advisable even in the case when considerations of engineering efficiency demand the utilization of a single channel for various purposes. This is a usual practice, for instance, in data transmission where various control signals are channelled via the same hardware, only to be redivided "at the receiver end". Let us assume that the vertex u possesses input ports x_i^1, $1 = 1, \ldots, N_i^1$ if there are N arcs in the system coming into u. Output ports y_i^x, $K = 1, \ldots, N_i^2$ are similarly defined. This means that each arc in the graph should be properly designated by a four index letter: d_{ij}^{lk}.

Various flows between different system components may be of discrete or continuous nature. For computer simulation every continuous flow should be converted into discrete (digital) form. A flow component (in a certain conventional sense) is assumed to be a quantum (which is generally different for different links) of the data transmitted, or a signal. Each arc s is characterized by its capacity M_d, and a signal s passing along the arc - by its transmission time τ_s and parameters $\{m_{sh}\} = m_s$ which are used (specifically, over a family of signals) to check the M_d constraint.

Employing this representation procedure is not always a cut-and-dried matter, especially when the simulated system is sufficiently complicated. It is important, however, that a model should come off isomorphic in structure to the system itself (at a certain level of aggregation). There is a variety of specific devices which enable this representation scheme to be used rather extensively. For instance, if the state of the arc is essential for the decision-making (functioning) of a certain unit, a dummy vertex linked to that unit is introduced into the arc. Another example: let us assume that some object of the system to be modelled is mobile (at a chosen level of aggregation) and may travel along a certain highway. If objects are so complex that individualized descriptions are necessary (even if completely identical), each of them is to be represented by a separate vertex, while a certain "controller" (represented by another vertex connected to each of them and characterizing the state of the highway) is to control and coordinate the movements. If no individuation is needed, the traffic is described by a flow along the arc with specific characteristics, whereas the role of the "controller" (much simpler than in the previous case) is passed on to one of the terminal vertices.

One of the most important parameters characterizing a signal is the time needed for its transmission along the arc from one of its vertices to another (sometimes we will refer to this as *delay time*). To describe the processes within a system one should choose a time unit Δt of the discrete time t. The magnitude of Δt determines the level of system aggregation "in time". It should be consistent with the level of "spatial" aggregation (in terms of components, products, production factors, etc.). Low values of Δt, i.e. a low level of aggregation (disaggregated description), correspond to the high dimensionality of the model, and not only in quantitative terms (i.e. with respect to dimensionality of vectors and matrices involved, etc.), but also qualitatively, since a greater number of different low-level components, events and relations are to be considered.

In choosing the Δt value, three basic considerations should be borne in mind. First, it should make the representation of real continuous time processes adequate (with respect to the modelling objectives). Second, a totally justifiable desire to reduce the model dimension (which in the final analysis will save not only computer time and data pre-editing costs, but programming efforts as well) should not result in an unreasonable reduction of the aggregation level. Third, in signal modelling, the closed loops should be avoided. The latter requirement stems from the fact that an increase of Δt generally means an overall transition to a higher level of aggregation, and as such, is usually accompanied by lumping together several entities of the lower level of aggregation. The result is that several causally related events may be merged into one, and thus in some cases a closed-loop chain of cause-effect relations may fall into a single time interval and become lost. According to the first and third requirements Δt may be chosen as a common denominator for time intervals of adjacent vertices interaction. However, this should not mean that one must try and choose Δt as high as possible, for it will violate the second requirement. The resulting discrete model of a continuous process may prove to be excessively crude, and consequently lose its utility. And of course, along with the processes of vertex interaction one has to also allow for the processes occurring in each of the vertices considered, and make the results of this scrutiny influence the selection of the final value for Δt.

Simulation of Systems with Changing Structure

Let us refer to a time unit Δt as a *step* of the simulation system operation.*

In passing from one step to the next, the state of the vertex may change under the influence of input signals and internal processes. Let us assume that the number of such states is infinite, but each may be represented by a finite set of parameters. Let t be the discrete time (for simplicity let $\Delta t = 1$), $P_i(t)$, the internal state of the vertex u_i at time t; $S_i(t)$, the total input signal applied to all input ports u_i; $F_i(t)$, the total output signal generated. In a general case, the evolution of u_i is described by the equations $F_i(t + 1) = \Psi_i(P_i(t))$, $P_i(t + 1) = \Phi_i(P_i(t), s_i(t + 1))$.

Thus the simulation model functions through a sequence of steps, each comprising three stages:

(1) Generation (for all i) of output signals F_i by the vertex u_i depending on its current state (before it is altered) in accordance with Ψ_i transformation,

(2) formation of input signals S_i for each vertex u_i from the ensemble of output signals as generated by the remaining vertices,

(3) transition of each vertex u_i into a new state determined by its former state and the total input signal in accordance with the function Φ_i.

It is the second stage of the sequence which ensures the model functioning during one step as an integral, systemic combination of actions of its individual components.

DESIGN OF A SIMULATION MODEL AS A SYSTEM PROGRAMMING

As already noted, in the simulation of a complex socio-economic system, even if its functioning is studied only under the conditions of the invariant structure, the information utilized in a simulation model may evolve in time. Thus, one has to find a storage technique to accommodate this change. A possible solution is to employ some conventional data structure representation techniques, allowing for their evolution.

There are several data base management techniques which provide for at least group relations [4]. A *group* is an ordered or unordered list whose every item contains a data set describing its instance. Some of the group instances may be groups in their turn. The group structure enables new items (vertices) to be added or existing ones deleted at will. The description of the object structure by elementary data and group relations is contained in the so-called scheme. The scheme describes a *file*† as a whole, whose

---

*Simulation models with a fixed step are used here only for simplicity; the approach considered is also valid for other cases, specifically, in so-called "asynchronous" simulation when each new step of operation is initiated by some specific event rather than by the passage of time.

†A file is a properly organized set of data regarded as an entire whole and provided with a unique name to make its retrieval possible.

structure represents the object functioning and relations between the objects. However, quite a few semantic relations are not expressed by this structure and are taken into account at another level of representation, through the algorithms describing the system and object functioning. Moreover, some relations do not find an adequate expression on this level either, and are accounted for by a system's programmer directly during his work with a model.

At the moment, two pattern-oriented data processing approaches prevail. The first [1] requires the development of an independent data base management system providing for storage updating and data retrieval (usually implemented through an interpretive routine) with a consecutive reading of the necessary item values into the operating memory space especially allocated for the process. As for the model itself, in this approach it is represented in one of the standard programming languages (FORTRAN, ASSEMBLER, etc.). The second approach (as represented by [9]) relies on the development of a sufficiently rich problem-oriented language whose sentences may deal with the objects of the required nature directly (database entries, structural expressions, etc.), and specifically with the problems of database updating and processing. However, a precompiler would be required in this case to translate the texts from such a language into some host language (e.g. ASSEMBLER). For instance, an ALGOL-like statement.

for A do B

is convenient when dealing with groups, A being the group name and B being the set of operations to be performed. By using this statement, one may implement a consecutive retrieval of the appropriate group instances along with performing over them the required transformations as described in B. The operator acts until the group runs out ot its instances, while the number of these instances does not have to be explicitly specified. A similar record processing capability is offered by the expression

RC (NXT (A) B)

where A is the group name, B is a set of operators, NXT is a function of transition to the following group instance, RC is a function of performing the bracketed sentence as long as the actions implied by it are feasible.

Using the notions and definitions introduced, a method may be developed for designing and operating a simulation system.

Preparatory Stage

(1) Identification of system components (vertices) and their interactions (connecting arcs); classification of the components and arcs by structure and functioning; definition of the schema (file label) for each of the components; development of algorithms describing the functioning of components and the system as a whole.

(2) The records, i.e. data carriers during the simulation, are divided into four categories. Records of the first category are used for component data files generation and (if necessary) their updating by a human operator (simulation system administrator) during functioning: addition of new objects or relations, modification of capacities, etc. The number of different records of this type depends on the structure of the system components. Output signals for the system under operation are recorded in the document of the second category, which after the analysis of constraints immanent for

the system are transformed into records of the third category, keeping track
of the state transitions of various system components at the appropriate
step of operation and records of the fourth category containing the output
information for the components at the subsequent steps.

(3) Initialization of component files which together constitute the database of the simulation system. This is done either by accumulating the data
for the records of the first category and filing it in appropriate files (in
accordance with file definition schema) or by the minor modifications
(usually simplification) of the respective information system used as a basis
for a simulation system.

Simulation System Operation

(4) Maintenance of the records of the first category and accumulation of
data recorded in the record file (if necessary).

(5) Generation of the records of the second category and their accumulation
in the same file.

(6) Checking for system constraints and generation of the records of third
and fourth categories using records of the second category.

(7) Sorting records by the required number of parameters in accordance with
the schema of database files.

(8) Database updating in accordance with data contained in records of the
first and third categories, and those records of the fourth for which the
delay is equal to one step.

(9) Record file updating. At this state the records listed in (8) are
deleted, and the delay time of the rest of the records of the fourth
category decreases by a unity.

(10) Return to (4) the required number of times or when some conditional
criterion is met.

Running through items (4)-(9) forms a single iteration of the simulation
process. When addressing a simulation system, a copy must be made of the
database part that is to be used as the database of the simulation model.

MODIFICATION OF THE SIMULATION MODEL STRUCTURE

Let us assume that graph vertices are divided into N classes, each associated
with its own data file. For convenience they must be merged into a macrofile
introducing into a structural description of a model at a new higher level
and new attribute for identification of original files. Each of the
original files will be associated in a macrofile with a group GFL_i,
$i = 1, ..., N$, whose structure is left intact, and if in the original file FL_i
each vertex has been associated with an item, in a macrofile an instance of
group GFL_i is.

Let the arcs connecting the vertices be divided into K classes. Group GFL_i
will naturally contain GX_j, GY_i groups (in general $j, l = 1, ..., K$),
associated with the classes of incoming and outgoing arcs of the given
vertex, i.e. its input and output ports.

In grouping vertices and arcs into classes, we reveal their semantic similarity expressed in their equivalence with respect to the modes of processing, i.e. all system algorithms act similarly on representatives of the same class. All exceptions, if any, are built into the algorithm itself. The number of exceptions demanding substantial individual processing should naturally be low, otherwise the concept of the class becomes meaningless.

Let us initially consider the problem of the addition of another component which is not conceptually new, i.e. which falls naturally into one of the classes formed earlier. An algorithm describing functioning can deal with the instance of the group GFL_i corresponding to that component, since the structure of the instance is standard.

Let us next consider the performance of the algorithm controlling data processing in respect to components connected to the newly added components. The algorithm remains unchanged with new instances of groups GX_i and GY_j added. And this is the way to provide connections with a new vertex. Semantic rules of component functioning which have not been explicitly expressed in the algorithm should, however, be observed. This is why in creating a new output arc for the component u_i a new instance of group GY_i has to be generated in GFL; and for the file associated with some component connected with u_i by the given arc new instances of GX_i groups must be generated and the already existing instances of GX and GX_i groups updated (if necessary).

If parameters describing the connection being considered are time-varying, the same arc may be associated with the instances of several groups. In this case, the original group of data representing the connection is decomposed into several groups containing parameters equally dependent on time.

The addition of a new input arc is done similarly. The remaining two system algorithms - the first testing signal acceptability, and the second transforming the signal - should meet the sole requirement: to be invariant to the number of components in each class. This requirement is obviously not especially restrictive.

Thus, if a system satisfies the assumptions mentioned above, the addition of one or several components of the type associated with classes formed earlier would only require filing new records in the file and making corresponding changes in the groups describing component connections. All these modifications are provided by the information system automatically using data on class affiliation and the quantitative characteristics of the component added.

The addition of a new component in the case when it cannot be placed into any class formed earlier would naturally involve software modification. The amount of effort by the programmer to do so may be treated as a quantitative evaluation of the "novelty" of the component added. This type of change is indispensable for a description of the component's inner functioning. It might also be necessary when describing new component relations with already existing components of the system, and in the modification of the algorithm for testing signal acceptability. However, the addition of a new component demands only local changes in program, but not the restructuring of the simulation model as a whole, whereas a simulation system in accordance with the definitions takes care of all algorithm modifications required. It introduces a new class of components, new classes of arcs, if necessary, thus reducing the problem to the previous case. Naturally, the programmer has to express the implied modifications of algorithm in compliance with simulation systems requirements. The richer the system and the means of systems programming employed, the less stringent the requirements.

The elimination of one of the objects which previously existed in the system does not present any new methodological problems, and is carried out as above.

In the lifetime of a simulation system, it may become necessary to study a structure that is conceptually different from the one provided for in it. It may happen that in this situation, a new simulation system will be needed, since the required "automated" modification of the initial simulation model is either impossible or prohibitive in terms of programming costs. This, for instance, may result from the artificial nature of the problem of adapting to the requirements of the available simulation system which the programmer may have to face.

The approach described may offer interesting possibilites for the development of simulation systems modelling real systems with a varying structure. There is every reason to believe that it will help minimize the trouble of introducing into the model changes reflecting structural variations in the simulation system. As a result, simulation may become much more efficient in terms of the time needed, which in turn will greatly increase the efficiency of planning and forecasting. Equally important is the resultant economizing of costs and development time, since for a wide range of problems there would be no need to develop the simulation system from scratch each time spontaneous (or designed) changes occur in the real world, which may be represented simply as changes in the system structure. Moreover, systematic development of simulation systems will help identify some aspects common to these systems in general, and will thus facilitate their design for the new type of reality.

REFERENCES

1. T. T. Arkhipova, V. A. Roshchin and I. V. Sergienko, On the solution of the class of processing organization problems in computer data processing systems, *Kibernetika*, 118-123 (1973), in Russian.
2. N. P. Buslenko, Large-scale Systems. Moscow, Nauka (1968), in Russian.
3. N. P. Buslenko, Large-scale systems and simulation modelling, *Kibernetika* 50-59 (1976), in Russian.
4. *Feature Analysis of Generalized Data Base Management Systems*. CODASYL Systems Committee Technical Report (1971).
5. Yu. A. Gastev, *Homomorphisms and Models. Logico-algebraic Aspects of Modelling*. Moscow, Nauka (1975), in Russian.
6. A. Ollongren, *Definition of Programming Languages by Interpreting Automata*. New York, Academic Press (1974).
7. K. Jacobs (editor), *Selecta Mathematica*, Vol. $\underline{2}$. Berlin, Springer (1970).
8. I. L. Tolmachev, Certain aspects of computer data organization and processing, *Ekonomika i Mat. Methody* $\underline{13}$, 543-549 (1977), in Russian.
9. V. A. Uspensky, *Lectures on Computable Functions*. Moscow, Fizmatgis (1960), in Russian.

METHODOLOGICAL ASPECTS OF PRACTICAL APPLICATIONS OF SYSTEMS ANALYSIS

O. I. Larichev

The ideas of the systemic integrated study of complex problems are gaining increasing recognition. Systems analysis can be regarded as a practical tool for implementation of these ideas. Many recently published textbooks and papers (e.g. [19] and [25]) deal with it or with its successful application in managerial environments (e.g. [18]). At the same time a fair number of publications appearing in the past decade indulge in criticizing some abortive systems analysis applications [9, 21]. The aim of this paper is to examine systems analysis procedures and postulates from the point of view of their relevance to a wide range of applied ill-structured problems [17].

SYSTEMS APPROACH

The words "system", "systems", "systemic", etc., are used in different expressions and combinations. In the pertinent literature, systems engineering [6], systems programming [3], systemic organizational design [28] and many similar terms are used along with the systems analysis notion. The main concept of system in its current meaning undoubtedly stems from the ideas of general systems theory [5] and cybernetics [27].

In the general framework of this overall systems approach, many different directions of study are feasible. For instance, one may regard it from the philosophical and methodological points of view. We do understand the importance and diversity of these various aspects of systems studies. But in this paper we will limit ourselves to consideration of just one of them, reflecting only the pragmatic side of systems analysis instructions.

The main concepts of systems approach are "system", "process", "input", "output", "feedback", "constraints", etc., [16]. These concepts have proved their validity in the analysis of systems of diverse nature. As for ourselves, we are going to look at them as they emerge in the study of the process of choice of a unique installation project or the process of compiling a departmental five-year plan, for instance. The analysis of these processes makes it possible to identify various systems dealing with similar problems (and the sub-systems they consist of), to establish their interdependence with different systems, and to identify their inputs (input information), outputs (decisions), feedbacks (decision analysis and verification) and constraints (resources, labour, etc.).

But what is the usual meaning of "systems approach" in this context? To answer this question, the existing recommendations as to how to tackle problems of a diverse nature in a "systemic" manner need to be examined. For instance, systems engineering recognizes [6] the following main stages of problem-solving: problem diagnosis and choice of objectives, enumeration or invention of options, their analysis, the search for the best option, and finally, solution presentation. Systems analysis and operations research usually distinguish five main stages of development of a similar scope: objective (or objectives) formations, search for the options consistent with the objectives established, evaluation of the resources needed for each option, model development, and the discovery of preference criteria suitable for the ultimate choice. In this context, the principal difference between systems analysis and operations research is that operations research relies on mathematical modelling, while systems analysis mainly uses logical models interrelating systems objectives, action options, environment and resources. The main stages of development referred to above may be compared to similar stages of managerial decision-making, which are [28]: establishing organization objectives, identifying the problems to be overcome to attain these objectives, analysing these problems and providing an ultimate diagnosis, searching for the solution, evaluating different options and choosing the best of them, discussing resulting conclusions with management and submitting them for final approval, finding the way to implement the decision taken, managing this implementation phase and validating the outcomes.

Similar stages of development are mentioned in various textbooks and papers - probably everywhere where authors endeavour to deal with complex problems in a systematic way (compare the general postulates for inventors looking for creative solutions to fundamentally novel problems [1]).

It is this precise structuring of research activities, this regard for objectives and the costs of their attainment, this identification and systematic evaluation of various options and this insistence on presenting the rationale for the choice that constitute the common features of the many papers dealing with systems approach. These ideas are of such a general nature that it is only fair to assume that many rationally minded people have used the similar structuring of their own complex problem solving processes for quite a long time.

Thus a systems approach to various problems primarily is manifested by the dissection of the system from its environment and by provision of an organized logical pattern of problem analysis. We can call this pattern the general systems approach schema. But in this regard, what makes the various modifications of the systems approach to dissimilar problems so different? First of all there is a difference in the way the analytical comparison of options is made. For instance, systems engineering makes use of standard evaluation techniques suitable for various classes of engineering systems (electronic circuitry, automatic control systems, etc.). Operations research relies heavily on several established techniques (mathematical programming, probability techniques, network analysis, etc.). As far as systems analysis is concerned, the main role is played by the "cost-effectiveness" analysis technique.

In dealing with a certain complex problem, our general systems approach schema is sometimes used without any reference to any special means of analytical comparison of options. This kind of approach has come to be immensely popular recently. It is now even hard to find a complex problem where a similar approach has not been recommended. Many attempts to directly apply the general systems schema to the solution of various problems are described elsewhere. For instance, one of the most detailed descriptions of this type is offered by Young [28]. He specifically analysed the process of a new hospital management system design. Meanwhile neither this nor any other similar presentation contains any conclusive proof that the general systems approach schema is a useful instrument of problem solving by itself. Of course, it is always useful to ponder the aims and objectives of your actions and to match them with the available resources, but there is nothing very new about it. In complex practical situations, similar postulates do not contribute very much. A meaningful analysed demarcation of system borderlines is always a result of creative thinking. And the apparent general purposefulness of the established analysis pattern breaks down completely when confronted with the complex ill-structured problems of real life. Identification of objectives and problems often depends on the general idea of the solution, i.e. on the choice of certain options, whereas the decision rule itself depends on the available set of options.

In general, the situation looks like this: logical structuring and systematization of the various stages of complex-choice problem solving might be quite expedient for some managers and consultants. But they do not offer a blank cheque for problem solving. Young claims that systems approach offers itself as an engineering technique. Unfortunately, it has still a long way to go before it reaches this stage of maturity. To be considered a constructive means of problem solving, it has to develop some specific set of analytical tools, some specific techniques for the evaluation and comparison of options.

Evidently systems approach application, even at the level it is currently capable of, may be beneficial. The damage stems from its excessive overevaluation, from the resulting fetishism, and from the fact that it has become the vogue of research. Unfortunately there is now a tendency to use the term "system" indiscriminately. To quote Gvishiani [7]: "The term 'system analysis' has become very fashionable and in this capacity is tagged on every managerial action which has proved to be successful."

This fetishism becomes exceptionally dangerous when not engineering systems, but rather social (transport, urban, managerial and so on) ones are being dealt with.

This reliance on the omnipotence of systems approach makes certain scientists quite confident that knowledge of the general pattern of investigation alone is sufficient to successfully deal with the various problems arising from different ways of life. However, in different systems even the concept of rationality is vastly different. For instance, rationality criteria valid for health services might not be so in public transit systems.

Thus, the pragmatic potential of the general schemata of systems approach (i.e. the list of the stages of a typical research inquiry) is relatively moderate. But at the same time, the didactic impact of systems analysis ideas is extremely important. These ideas should find their place in the curriculum of universities, vocational schools and just ordinary schools. The ideas of a systemic, consecutive and consistent, multistage approach to complex problem solving may be quite beneficial for intelligence shaping and development.

OPTIONS COMPARISON TECHNIQUES

Recognition of the existence of the general systems approach schema makes it possible to consider systems analysis as a combination of this schemata along with the correct specific options comparison technique. But what technique are we referring to?

Systems analysis is generally considered a tool for dealing with ill-structured problems where qualitative, uncertain and ill-defined aspects prevail. To see what distinguishes these problems from others, we may compare them to the problems dealt with typically in the general framework of operations research [26]. This options comparison is organized in the following manner: first, one has to develop a mathematical model that objectively reflects the principal features of the problem being investigated. Then one has to define a proper criterion that represents in an unambiguous and explicit manner the requirements to be met. The best choice then corresponds to the optimal value of the criterion. However, in this case the problems are well structured.

In contrast, in ill-structured problems the lack of data needed for the development of an objective model is typically assumed. It is precisely because of this insufficiency of objective data that we are unable to weave different aspects of the problem together into a single integrated model. Instead we are forced to consider them on their own as separate evaluation criteria, and the problem itself turns out to be a multicriterial one. The data needed for matching the different criteria and for options evaluation in accordance with the different criteria should be provided by a decision-maker (DM) and his experts. As is correctly noted in [23], systems analysis is tailor-made for those problems where the trade-off between different criteria is made on a judgemental basis.

Of the different modes of options comparison in the systems analysis framework, the most important is the "cost-effectiveness" technique. It provides specific models of cost and effectiveness evaluation, and relates to each option several values obtained through their use. The ultimate decision rule (i.e. trade-off between the cost and effectiveness values) in certain cases employs the value judgement data of the DM. Three main approaches to cost-effectiveness aggregation have been singled out. The first requires minimizing the cost of the options, with an effectiveness no lower than a certain level (i.e. to choose the "cheapest" alternative providing a predetermined benefit). The second approach demands the maximization of effectiveness while keeping the cost no higher than prescribed (this is a case of budget constraints), and the third demands the maximum ratio of effectiveness to costs.

The rationale of the first two approaches is quite obvious. In essence, they reduce one of the criteria to a role of a constraint because it would be absurd to demand the simultaneous maximization of effectiveness and minimization of the cost. As to the third approach, Hewston and Ogawa [8] warn against the liberal mechanistic use of it, since one may obtain the same value of ratio for quite different combinations of effectiveness and cost values.

The first attempts to apply the "cost-effectiveness" technique were made
within the domain of a military systems analysis [25]. The problems with
which one had to deal there were of the borderline type between well- and
ill-structured ones. These problems also provided the opportunity to develop
objective cost and effectiveness models. Unfortunately further attempts to
follow a similar pattern when dealing with genuinely ill-structured problems
found these opportunities to be lacking. As a result, multiattribute or
multicriteria choice models emerge, allowing each option to be represented
by an entire vector of evaluation results.

At present, there is a variety of different multiattribute evaluation and
comparison techniques that are often referred to as decision-making
techniques. In all of them, these are experts (sometimes DMs) who evaluate
the options in accordance with different criteria. What makes them different
is the ways they transform these evaluation results in an integrated utility
evaluation. These techniques may be classified as follows:

In the first group (direct techniques), one can stipulate those techniques
which a priori assign the dependence of option integrated utility value to
partial evaluation values. This dependence is chosen in the form of a
weighted sum of partial evaluation results, with the weights representing
the relative importance of different aspects or attributes [4]. Decision
trees [20] may serve as another example of direct technique. In accordance
with this technique, one has to consider each decision (for instance, to
build a plant according to policy A or B, to construct its shops according
to technological design or D, etc.) one after another, generating different
options for the final choice from the different sequences. After that, an
appropriate rule makes it possible to calculate for each option its success
probability, which is to be multiplied by the option utility (in a monetary
form).

The techniques of the second group (compensation techniques [15]) compensate
differences in the partial evaluation of a pair of options, and thus make it
possible to find which of the two is preferable. This approach is
conceptually the simplest. According to it, one simply has to list all the
advantages and shortcomings of every option, discard all the advantages or
disadvantages common to each, and study only those which make the options
different.

The main instrument of the techniques of the third group (incomparability
threshold techniques [22]) is the binary relation among different options
showing which of the two options being considered is preferable (by a
majority of criteria). By checking this relation, we can sort every option
pair according to its comparability (when either option may be preferable
to the other, or may be indifferent) or incomparability and by varying the
basic relation may obtain a number of comparable pairs.

The fourth group (axiomatic techniques [10]) can be said to comprise those
which from the outset postulate several properties of utility function
representing the integrated preferences of the DM. Information obtained
from the DM in this case is used mainly to check the validity of these
assumptions, among which the most popular is the assumption of independence.

In the instance when at least a partial model of the problem under
consideration is available and may be programmed for a computer, one may
adopt one of the techniques of the fifth group (man-machine decision
systems) [13]. Here the desired interrelations between different criteria
are revealed through man-machine interactions.

Existing decision-making techniques are reviewed in detail in the survey [11].

Although the focal element of each of the techniques mentioned above is the use of information obtained from DM and his experts, more often than not the crucial question of how to acquire this information is largely overlooked and at times completely neglected. Sometimes DMs are immediately asked to select a choice model, an aggregation technique which may be sooner asked than done, sometimes they have to provide direct estimates or offer value judgements on multiattribute comparison, and sometimes to assess numerically judgemental probabilities.

Of course, different problems require consideration of different sets of options. For the problem where an alternative dominating all other (or nearly all), criteria exists, and where one has simply to identify the best possible choice, every existing decision-making technique is equally valid and will lead to the correct result. Unfortunately however, there are plenty of problems of a far more complicated nature, and for them decision-making is a long way from being that simple. But how to choose a proper decision-making technique for a realistically complex problem?

In my opinion, this choice should be based on the extent to which the requirements for information from the DM inherent in the technique harmonize with the existing accessibility and reliability of this information. To be able to do this, one has to use the results of specific psychometric studies (e.g. see [24]) and to classify different tasks in accordance with their complexity for the DM. This will provide one with a solid foundation for a rough comparison of different groups of decision-making techniques. However, even now one may assume that comparability threshold and trade-off techniques are more reliable than others. Of course, whenever possible (i.e. whenever a partial model of the process under scrutiny is available), it is desirable to select most of those man-machine decision-making techniques which do not demand direct multiattribute option comparison. Axiomatic and certain direct techniques should be used with the utmost prudence, as they are the most demanding as regards the difficulty which a DM would have when trying to provide the necessary judgemental information. That is why the practical feasibility of these techniques is uncertain.

FEASIBLE RESEARCH DIRECTIONS

In practical systems analysis applications, the general systems approach schema and option comparison techniques merge in an integrated whole. It is no coincidence that both the advocates [18] and most ardent opponents [9] of systems analysis alike have firmly linked it to the "cost-effectiveness" technique. Therefore, the further improvement of analytical tools of option evaluation and comparison is of vital importance for future applications of systems analysis.

To successfully deal with ill-structured problems, one first has to master the sophisticated art of problem analysis of identifying objectives and means, and of comprehending the inner mechanics of the processes involved. The only way to acquire this mastery is through case studies. Nonetheless, at the same time new ways of multiattribute options comparison have to be developed. And in our opinion, they have to be based on considerations of existing abilities to obtain information from the DM and his experts, and to check it [12, 14]. These techniques should become a convenient vehicle for expressing their preferences.

Analytical tools making the task of managers faced with a complicated decision-making problem easier have been changing rapidly in recent decades. Systems analysis should not be regarded as something already finite. It is being developed parallel to the necessity of dealing with increasingly complicated situations. But even now it may certainly provide real assistance to those confronted with difficult practical problems and who are well versed in the art of problem analysis, and help them to improve the quality of their decision-making.

REFERENCES

1. G. S. Altshuler, *Invention Algorithm*. Moscow, Moskovski Rabotchiy (1973), in Russian.
2. I. V. Blauberg and E. G. Yudin, *Emergence and Essence of Systems Approach*. (1973), in Russian.
3. D. I. Cleland and W. R. King, *Systems Analysis and Project Management*. New York, McGraw-Hill (1968).
4. W. Edwards, *How to Use Multiattribute Utility Measurement for Social Decision-Making*. Social Science Research Institute of University of South California. Research Report 76-3 (1976).
5. *General Systems*. Moscow, Progress (1969), in Russian.
6. H. H. Good and R. E. Machol, *Systems Engineering*. New York, McGraw-Hill (1957).
7. D. M. Gvishiani, *Organization and Management*. Moscow, Progress (1972).
8. M. C. Hewston and Y. Ogawa, Observations on the theoretic basis of cost-effectiveness, *Operation Research* 14, No. 2 (1966).
9. I. R. Hoos, *Systems Analysis in Public Policy*. California, University of California Press (1972).
10. R. Keeney and H. Raiffa, *Decisions with Multiple Objectives: Preferences and Value Trade-offs*. New York, John Wiley (1976).
11. O. I. Laritchev, Multicriterial option evaluation techniques, in: *Multicriterial Choice in Ill-structured Problems*. Moscow, Institute for System Studies Press (1978), in Russian.
12. O. I. Laritchev, Systems analysis: problems and prospects, *Automation and Remote Control* 36, No. 2 (1975).
13. O. I. Laritchev, Man-machine decision-making (a review), *Automation and Remote Control* 32, No. 12 (1971).
14. O. I. Laritchev, V. S. Boytchenko, E. M. Moshkovitch and L. P. Sheptalova, *Hierarchical Techniques for Programming Scientific Research*. (Preprint), Moscow, Institute for Systems Studies Press (1978), in Russian.
15. K. R. McCrimmon and I. K. Siu, Making trade-offs, *Decision Sciences* 5, No. 5 (1974).
16. S. P. Nikanorov, Systems analysis: a stage of the U.S.A. problem-solving methodology development. An introduction to the Russian translation of S. Optner's *Systems Approach in Planning and Control*. Moscow, Soviet Radio (1969), in Russian.
17. S. Optner, *Systems Approach in Planning and Control*. Moscow, Soviet Radio (1969), in Russian.
18. D. Novick (editor), *Programme Budgeting*. Cambridge, Mass., Harvard University Press (1965).
19. E. S. Quade, Systems analysis techniques for planning-programming-budgeting, in: *Planning-Programming-Budgeting: a Systems Approach to Management*, Chicago (1969).
20. H. Raiffa, *Decision Analysis*, Reading, Mass., Addison-Wesley (1969).
21. H. W. Rittel and M. M. Webber, Dilemmas in a general theory of planning, *Policy Sciences* 4, No. 4 (1973).

22. B. Roy, Classement et chois en presence de points de vue multiples (le methode ELECTRE), *RIRO* 2, No. 8 (1968).
23. I. R. Schlesinger, Quantitative analysis and national security, *World Politics* 15, No. 2 (1963).
24. P. Slovic, B. Fischhoff and S. Lichtenstein, Behavioural decision theory, *Annual Psychological Review* 28 (1977).
25. E. S. Quade and W. I. Boucher (editors), *Systems Analysis and Policy Planning. Applications in Defence*. New York, Elsevier (1968).
26. H. M. Wagner, *Principles of Operations Research with Application to Managerial Decisions*. Englewood-Cliffs, Prentice-Hall (1969).
27. N. Wiener. *Cybernetics or Control and Communication in the Animal and the Machine*. New York, John Wiley (1948).
28. S. Young, *Management: A Systems Analysis*. Glenview, Illinois, Scott, Foresman & Co. (1966).

A SYSTEMS DESCRIPTION OF GOAL-SEEKING HUMAN BEHAVIOUR

N. F. Naumova

The systems study of goal-seeking human behaviour is one of those fundamental interdisciplinary problems in the humanities where the holistic representation of a subject-matter is a major challenge. Conventional approaches which recognize wholeness only with reference to entities of physical reality are not the most fruitful. There seem to be far more extensive possibilities in regarding the problem of wholeness as a basically methodological one and dealing not with an object itself but rather with the knowledge about it [9, 10]. This conception of wholeness is addressed not to integral objects themselves but to those cognitive situations which arise in studying them [10, p. 23].

Furthermore, in this approach the ideas of wholeness are used less for determining actual knowledge than its incompleteness, i.e. the gap between what has already become known and what is still to be revealed.

This "potential" layer of knowledge as an element of the "new" integrity may obviously concern not only the future, "lost" with rejected research alternative. The more "choice situations" (and consequently the more research alternatives rejected) there were in the history of a scientific discipline the more this possibility needs to be considered. However, to conceive an integral conception of the subject of study, of course, inadequate simply to allow for a return to these prior research alternatives; a way has also to be found to recapitulate and actualize them. For instance, this may entail returning to the stages preceding the development or the emergence of the present integral conception. Methodologically, it implies that the aim of the critical analysis should be not empirical or theoretical elements of the given conception (i.e. how "adequate" or "true" they are), but rather its fundamental (first of all, philosophical) assumptions. In my opinion all these observations apply fairly well to the current situation in goal-seeking behaviour research and define the main problem that a systems approach has to tackle.

ANALYSIS OF THE EXISTING SYSTEMS MODEL OF GOAL-SEEKING BEHAVIOUR

Goal-seeking behaviour (activity, action)* regarded as a system, or an integrity, may be characterized by the following:

(1) It presupposes the possibility of recognizing "goals", "means" and "results" of this behaviour (this decomposition applicable to the nature of any direct human behaviour),

(2) It is always related to some conceivable "clear-cut" situation, and because of this, is oriented towards some specific ultimately attainable, though restricted within the framework of the situation, goal (the specificity of the goal depends not on its "scale" but on the possibility of measuring the progress towards it, to register the degree of its attainment),

(3) This goal is always apperceived and appropriated by the agent of the behaviour (which is indispensable if we want to later introduce subjective utilities and preferences),

(4) There is always a hierarchy of goals somehow ordered according to their desirability, their preferential stability (within certain temporal and situational framework), i.e. (a) it can always be said which of two goals has a higher priority, i.e. preferable to the other and (b) after one goal is attained it is superseded by the goal with the "next" highest priority (but not from the "outside"),

(5) Means are always subordinated instrumentally to the goals (the means, i.e. methods and ways, are chosen according to their efficiency in goal-attainment; the means and the process of goal-attainment itself are not considered self-rewarding),

(6) Means are substantially and semantically independent of goals (i.e. means should not be inherently related to goals and as a result their choice is determined initially not by the nature of the goal but rather by the conditions, circumstances, and possibilities of its attainment),

(7) Results, implication and "efficiency" of behaviour are accountable, i.e. may be ascertained and accounted for (an activity is judged by its efficiency, i.e. by its result; the activity result turns out to be its goal),

(8) Goal-seeking behaviour is universally fashioned by a general-purpose mechanism of "decision-making", i.e. evaluation of alternatives, outcome analysis and the choice of action according to the relative utility of the expected result (such a scheme is assumed to be valid both for the choice of the goals and of the means, i.e. the goal is rejected if its attainment involves "too much risk and/or effort"),

(9) The attainment of a goal is expected to be socially rewarding (the result itself is the reward),

(10) Rationality of behaviour is provided by the proper analysis of goals, means, results and their scheduling.

---

*The words are used interchangeably henceforth. I will consider in this study some principles providing the basis for several papers [2-4, 23, 56, 47, 12] on goal-seeking behaviour.

There is much to substantiate conception of human goal-seeking behaviour as it has been presented above. First, it reproduces to a fair extent the specific nature of both management activity itself and the basic domains of its application - economy and politics. This is why the model is sometimes called the model of the "economic human being" (*Homo aeconomicus*). Second, this mode or component of human behaviour is the most easily observed, measured, formalized, analysed and thus predicted. Third, it is easily controlled, since its basic components (goals, means, rewards) are at the same time the components of the system of influence, stimulation, education and management. It is this concept of human behaviour which the management system employs to shape the desirable behaviour pattern.

The model has three important advantages. First is its universality, because of which non-pathological human behaviour may be accounted for more or less authentically. Second - and this is especially relevant for the control and management science - is its formalizability, i.e. the possibility of applying the formalized techniques of description and analysis. The third advantage is important for sociology and philosophy: the model describes the mode (or the component) of social behaviour of a human being which readily reveals the mechanism of social and cultural creativity, or of the design of new stereotypes and guidelines of behaviour. But the difficulties encountered in applying this model to the solution of certain types of problems alert one to its inherently restricted nature.

(1) *Empirical limitations*. The modes (or aspects) of empirically observable human behaviour which do not fit the "goal-seeking-rationality" model and testify to its "non-goal-oriented" and "irrational" nature can be easily discovered and described. From the standpoint of the psychological characterization, this deviation is manifested in any impulsive and emotional behaviour [17, 46], as well as in the behaviour determined by the sphere of the subconscious and the unconscious [6]. If a goal is to be treated as something possessing psychological reality and not as a simple concept providing a tautological definition for one of the components of goal-seeking behaviour (a goal is something in respect to which the goal-seeking behaviour is rationalizable) it becomes obvious that not all behaviour is goal-seeking.

From the sociological viewpoint the same qualification is still valid in the case of behaviour determined by social stereotypes [33] or social relations, e.g. power relations, [37], collective behaviour [49] and morality [55]. Identifying the behaviour modes existing along with the "goal-seeking rational" behaviour in this way has provided Weber with a basis for his typology of social behaviour. A recent typology by Riesman conceived as the description of historically changing social characters also documented the existence of "non-goal-oriented" behaviour and the meaningful heterogeneity of goal-oriented behaviour.

In management theory, all deviations from the schema of goal-oriented behaviour as outlined above are usually ascribed to the non-optimality of the human behaviour [34]. However, when applied to behavioural social mechanisms and functions, this "non-optimality" is quite rational as it allows a human being to conserve the energy of "decision-making" for the solution of vital and atypical problems.

(2) *Control limitations* (which follow from empirical and may be considered as a special case of empirical limitations). The model of human behaviour as described above does not allow for its specifically social dimensions. It cannot portray certain important characteristics of this behaviour, i.e.

awareness of other people, influence of culture and the role of social comparison. Their effect, however, can dramatically change both the content and logic of individual goal orientation. Hence the schema presents a distorted picture of individual behaviour and social interaction in general. Even though a "goal-oriented-rational" model helps reveal the controllable components of the behaviour, it does not provide an adequate prediction of the behaviour as a whole since statistically speaking behaviour depends as much on the need to conform to social standards as it is dictated by the individual-conscious choice.

Successful forecasting of social behaviour initially demands the study of the social mechanisms of its determination - culture formation, standards and values development, deviating behaviour, etc. In other words, it makes it necessary to study the functioning of social stereotypes rather than the complex individual inner mechanism which cannot be used to infer impersonal social laws (see, e.g. [40]).

To present social interaction as a quasi-economic game, the subjects of that interaction as players, winning or losing, succeeding or not in their objectives and society - as the field of such a total game or its result would exceed the limits within which the described model of the human behaviour can be validated. Even Pareto, who tended to treat social behaviour in economic terms, considered it essential to distinguish two different mechanisms: that of utility of a community, when each individual seeks maximal satisfaction of his own interests; and that of utility for a community when the maximal benefit is sought not for individual but for the community as a whole [42].

(3) *Methodological limitations*. To capitalize on one of the basic advantages of the described model of social behaviour, i.e. its ability to be formalized, a certain axiomatics should be introduced. This is based on the assumption of the quantitative nature of goals, which serve as guidelines of human activity, from which the following important consequences may be derived. A goal and the process of its attainment can both be expressed or described quantitatively. In particular, the moment of the goal attainment can be determined quantitatively. Different goals may be compared quantitatively - one can always say which of the two goals is more "important", "attractive", "higher", etc. Because of this, they can be hierarchically subjugated and arranged into a stable (within certain limits) order of priorities.

The analysis shows that the axiomatics has a far narrower scope of relevance than is usually assumed [38]. As I see it, this is why the application of the logical and mathematical tools of the decision theory, theory of games, etc. is not always fruitful in describing human social behaviour.

(4) *Theoretical limitations* of the model result from the attempts to make it universal, i.e. applicable to the description of all empirically existing types (aspects) of behaviour. To do this the meaning of descriptive concepts has to be broadened sometimes beyond the limits where they actually do possess the explanatory power.

Whenever, for instance, one might choose to define the concept of "reward" not tautologically (reward is "something the human being strives to attain the goal for" or reward is "any result of rational behaviour") but rather meaningfully (as something definite and objectively different from anything which is not a reward), it becomes obvious that not all even goal-seeking behaviour, strives for a reward, or is elicited and motivated by it. To be able to retain the category of reward, its meaning has to be expanded to find room for something allegedly covering all unexplained behaviour. For instance, it is assumed to include "inner reward" (or supersublimation, superself actualization, etc.). But if we define "inner reward" as a phenomenon which can be observed, objectively defined and measured, an entire range of behaviour types will not conform again. On the contrary if it is not to be defined objectively, and is understood as an unobservable "inner state", a behavioural "intervening variable" or a certain theoretical construct, it cannot play the role of an independent, constructive component of the model. A similar "blurring" affects the concept of "rationality", "consciousness", and "goal" of the behaviour.

(5) *Existential limitations* of the model are analysed in great detail elsewhere, primarily in philosophical literature [7, 15, 41, 52]. It acknowledges three fundamental limitations of goal-seeking rational behaviour which demonstrates why it should not be considered "a universal exhaustive basis of human existence" [41, p. 528]. First, rationalization of the goal of that behaviour type falls outside the realm of that behaviour, as it belongs to the sphere of human ideals and values. Second, the very logic of that action enables a human being to competently deal with only a situation familiar to him, when he is well aware of the goals to be attained and can plan the employment of means the situation offers. Third, an "engineering" vision of means inherent to this conception of behaviour, i.e. given means with a certain self-sufficiency and evaluating them only by their efficiency rather than on their own merits, opens the way for the substitution of means for the goals, loss of goals and eventual depreciation of moral guidelines.

Two other features of goal-seeking behaviour revealing its restricted rationality are noteworthy.

The concept of activity based on the goal-orientation of behaviour leaves it destitute by depriving many of its components, spheres and periods of substance and meaning. Anything regarded as means automatically loses its independent meaning. The result is that this conception greatly narrows down the area of activity meaningfulness by subordinating some (perhaps of no little importance in their own right) realms of life to others thus relegating them to means. Human existence loses its continuity and becomes fragmentary and disjunct. Fuller and continuing experience is gained by eventually renouncing an instrumental attitude to the world, manifest in its conversion of everything into means.

Moreover, the conception of universal goal-seeking makes our existence unilinear and one-track, stripping it of its possibility of choice. "Decision-making" implies the rejection of many alternatives for the sake of one. The more decisions are taken, the more alternatives are rejected and every subsequent "consistent" decision (corroborating and reinforcing the preceding one) makes a return to rejected alternatives or to "overlooked" or "undiscovered" choice situations even less possible.

The existential narrow-mindedness of the conception of goal-seeking behaviour conception is demonstrated by and is the reason for the instability of interpretations revealed in the constant metamorphoses of its components, and its transformation into another type of behaviour. The transformation of means into goals and of goals into values; the rejection of external social rewards for an inner reward; the selection of means not on the criterion of efficiency but on individual acceptability or according to social standards, etc. are all inevitable components of goal-seeking behaviour.

All the arguments outlined above point to the following conclusions about the nature of the objections to the model being investigated. First of all, it might be pointed out that the model "misses" many things, and hence needs to be supplemented. But although the components unaccounted for (such as social mechanisms) may be quite important in their own right, this objection is in fact not a crucial one. One may enlarge the model by allowing for the missing components though of this kind appendage frequently happens to be on thin ice theoretically. For instance one may take social mechanisms of orientation into consideration by regarding them as normative constraints on an individual level. The second area of reservation involves these attempts of model generalization. Any attempt to incorporate "everything" into the model results in stretching quite a few thoeretical points, and some rather dangerously so. However, neither of the reservations at all involves that "potential" scope of the integrity of the ideas which is so badly lacking. Only the third group of reservations (which we have chosen to call "existential") prompts a search in this direction. Because of this the main problem is rather to explicate and rethink its most general assumptions than to generalize or refine the existing model using sociological and psychological data and theories. There are essentially two major premises of this kind, both of which are within the framework of that line of philosophical thought on goal-seeking behaviour which emerged last century. This trend is initially characterized by the fact that it considers the process of goal-assignment or goal-setting as occurring outside activity [41], and, second, by the unilinear interpretation of rationality. As disciplinary scientific research of human goal-seeking behaviour, this is revealed in the fact that the mechanisms of goal-setting are either completely neglected (the goal is dictated from outside and thus established in advance) or equated with the mechanisms of goal-attainment (i.e. goals are chosen in the same way the means are). Meanwhile the degree of rationality is inferred to be quantitatively measurable as a higher or lower degree of awareness, logical, consistency and behaviour conformity.

At the same time parallel to the above model of goal-seeking behaviour is also another, alternative approach [51, 57, 54] which treats goal-setting as not only a necessary but also a system-forming component of activity with a complex structure and specific mechanisms of functioning. It implies the existence of conceptually different "rationalities" whose nature depends on the specific type of goal-setting involved. According to this conception, goal-seeking behaviour can be considered as a system whose component and functioning depend primarily on the nature of goal-setting and on the interaction of its many mechanisms.

Here we are confronted with the problem of examining the mechanisms of goal-setting in relation to the goal-seeking activity of the human being as a whole.

MECHANISMS OF GOAL-SETTING (GOAL-ORIENTATION)

I will consider the pattern of the activity goals (in the broad sense of the expression) and of its guidelines along two major dimensions. One characterizes the orientation of human activity to the surrounding world or to the individual himself, his personality. The second takes into account the existential or normative nature of any guideline or, which in this particular case happens to be the same, restrained (i.e. prompted by natural requirement and available means) or unrestrained (prompted only by the concept of oughtness, of obligation) goal-setting.

(1) The first, most complicated mechanism of goal-setting is expressed by the fact that a human being always possesses in one form or another* a certain life "design", life plan [43, p. 373], life objection, ambition or aspiration [28, p. 220], aspiration project [7, p. 108], project [32, p. 556] or general motto of existence [16, p. 127]. This mechanism stems from the human being's ability and willingness to project his own life into the future not only for establishing specific goals to be pursued but also for self-design, i.e. not for the partial but for the complete extrapolation of himself onto the future, for finding a place for the future, for the possible, in his real life. This mechanism of goal-orientation functions in the sphere of personality content, i.e. those totally individualized, personalized meanings of actions and motives which do not have any objective superindividual sense and are at the same time the most profound characteristic of the human personality.

Three different mechanisms furnish the genesis and apprehension of such generalized integral guidelines of behaviour.

Because of the permanent preponderance of certain motives, human life (the inner state of the individual) spontaneously becomes goal-seeking, and the goals are essentially of an emotional nature.

As a result of constant inner development (which is basically self-cognition) a certain life objective (leading motive) emerges and stimulates the formation of hierarchical pattern of meaningful motives. The goal may collate all the meaningful elements of life in a single focus. However, "even when a human being has a clear-cut guiding line of life it cannot stay unique" [28, p. 221]. Through acts of conscious (though not necessarily rational) volition, a human being makes an unrestrained (i.e. determined by individually internal and not by external circumstances†) choice of self that is not the choice between superior and inferior empirical alternatives and specific goals, but rather of an individual way of life, the choice between good and evil in his own existence [19, 20, 32].

The interactions of these mechanisms result in the evolution of a life project or a life design, which possesses characteristic wholeness, dormancy and potentiality. The design is not structured, it does not break down into goal, means and result (reward), causes and effects, and should not be regarded as a rational "conception" of personal life. It is a specific, permanently transient field of ready (both implemented and potential) "solutions" and specific choices that are actualized in a respective situation. At each specific moment a person has already made a choice (for the future as well), and his job now is simply to find, "recollect" and comprehend a prefabricated decision. This is why the design-induced behaviour may seem "impulsive" and irrational because of the instantaneity of "decision-making" and lack of rational validation.

---

*The degree to which this existence is expressed depends on how well a human being is "provided for" in biological, psychological, power supply and social terms [6].

†By external circumstances I imply the possibility of realizing a chosen way of life, the availability of means required, the possible consequences, value and standard socially accepted ideas, etc. By internal circumstances I mean a tendency to keep one's own ego, "to resolve to dare".

Genetically, the design is determined by various other orientation mechanisms in many ways. In actual fact, a design is only a starting point for these mechanisms, though it is impossible to infer them from it. The mechanism's function is to form, preserve, and reproduce a personality, and thus the individuality of a human being. The project as a guideline of activity reflects both existential and normative components expressed through formal requirement to maintain the choice of self.

(2) The concept of obligation serves as another guideline of human behaviour: genuinely conscious and free activity should combine the cognition of the reality and the understanding of obligations [57, p. 176]. Morality is exactly this specifically volitional self-regulation of behaviour which compels the human being to follow not his inclinations (needs and aspirations), but his duty [25, p. 234]. In this sphere of morality a human being himself determines his own laws of behaviour (in this sense he is free) which is based not upon his nature (needs, interests, aspirations), but rather upon his own conception of obligation. Some of the characteristic features of the behaviour imposed by the concept of obligation are listed below [15, pp. 356-365].*

Moral commitment is a goal in itself, i.e. cannot be regarded as a means for anything else. Behaviour consistent with morality but induced by aspirations other than that of honouring the duty is, strictly speaking, not moral. Thus, an evaluation of one's own behaviour from the standpoint of its dueness should be considered a secondary phenomenon that does not express the essence of morality.

Moral commitment is imposed in an unconditionally obligatory fashion, i.e. regardless of possible outcome. It has nothing to do with an appraisal of a result or efficiency (from either the personal or social standpoint [31, p. 136]). The propriety of the moral commitment is not the same as the prerequisites of its feasibility or expedience. The essence of moral commitment is not the achievement of a result but in the aspiration itself, regardless of the consequences, which may ultimately prove undesirable. This is why in this instance no information on the situation, possible outcomes, etc. is required.

The unreserved obligatory nature of morality is expressed and implied in the requirement of absolute unselfishness, in the absence of any anticipation of reward ("moral satisfaction" included). Dueness must be understood as something entirely different from what is due to take place in future.

The described mode of behaviour characteristically combines the "goal" and means, and determines the choice of means by the content of obligation. In goal-seeking behaviour, a choice of this kind is dictated by efficiency considerations ("the end justifies the means"), whereas in moral behaviour, the ways to discharge a duty are always innately consistent with it (the means create the end).

Commitment to an obligation is a deliberate act of individual self-formation, orients behaviour in the context of maximal personal and/or social uncertainty, and opens the way for strategic behaviour exceeding the

---

*My description of moral behaviour is based on that of Drobnitsky whose position is totally different from the prevalent utilitarian concept in all its versions, and makes it possible to discriminate between "goal-seeking" and moral behaviour.

boundaries of the available limited situation as perceived and rationally recognized.

(3) The first two mechanisms (1) and (2) involve the orientation of a human being in the subjective world of his acts and aspirations, contributing meaning and valuation, whereas the third mechanism orients him in the sphere of objects in the outer (natural and social) environment. The objects are endowed with a value, i.e. a significance, or subjective relevance; they are related to human needs, interests and aspirations. Values are attached to any object, material or ideal, real or imaginable, to any idea or establishment, towards which the individuals "occupy a position of assessors, assign to it an important role in their lives, and feel the wish to possess it as a need" [48, p. 52].

Values are formed during a person's social activity, during his aspiration to satisfy his requirements and transform them constantly, choosing the priorities and the means under the influence of the social norms regulating these priorities and means. Nonetheless in due time they acquire independent significance as guidelines for human activity.

Once the process of values formation has been completed and values become wholly developed and established, they can be considered functionally irreplaceable entities which are useful and important in themselves, and not because they serve a certain purpose. There are so-called terminal, intrinsic, end and purposive values, and referring to them, one can speak of a loss or acquisition (but not of a "compensation"), of simultaneous existence (but not hierarchy).* In contrast, instrumental (operational) values may form a hierarchy since in some situations they can be a means of implementing purposive values, of concretizing them [58, 50, 26]. However, on the whole, values do not seem to form a stable system,† since the constant emergence of new and functionally independent values destroys their established order. Thus a "system" of values, by definition, harbours the roots of permanent self-destruction. It is organized and ordered from outside, by external (mainly resource) constraints which prevent a human being from seeking to actualize every value he holds and which make him constantly choose between them, to rank and coordinate them.

Value-oriented behaviour is exercised and revealed in the selection of variants which can be assessed neither on a rational basis (by calculation) nor by social stereotypes (social patterns). Their valuation does not arise from their usefulness or necessity, but depends on the particular conception of "good" and "bad". The values act and exist only at a moment of choice and assessment, or to be more precise, only when they reveal themselves as a component, a guideline of behaviour. Their function is to create an orderly and meaningful picture of the world. They provide the basis for preference formation and for the choice and assessment of variants, and to a certain degree restrain our actions, i.e. not only direct but control them as well.

---

*Empirically this was studied as the possibility of comparing terminal values. The results were ambiguous [50, 13, 14, 22, 45].

†Yadov and Kluckhohn have a different opinion.

(4) Social and cultural norms* (as well as values) help human beings find their way in an external (natural and social) world.

But if values express relations between subjective and objective worlds in the light of human needs, interests and aspirations, the major factor in the sphere of norms is social demand. These demands determine socially and naturally (technologically) permissible, acceptable, possible, expected, and approved behaviour. A norm is a rule, standard or mode of action which exists in the given society and is accepted by the given individual and rules his behaviour in the given situation. It expresses the socially approved established behaviour and delimits allowable range of actions, i.e. the boundaries within which the individual can seek alternate ways (means) of attaining his goals [36, 21, 44].

The substance of norms depends on the form the social activity, labour and social communication have acquired - in other words, on the substance of collective experience. They are evolved and clarified in the process of direct collective communication. The find validation in the nature of obligations which have been universally accepted and in the terminal values that exist. Once certain norms have become generalized, non-specific, profoundly rooted in general consciousness and acquire emotional-evaluative overtones, i.e. are expressed not in "permitted-prohibited" terms but rather in terms of "good and bad", they may take on the functions of values or obligations. Likewise, when connections between certain values and actual needs or aspirations become vague and these values reduce to just another consideration regulating our behaviour, they develop into norms and that is also true for the obligations imposed and controlled from outside.

Behaviour regulated by norms lacks inner structure. Within it one cannot differentiate between the goals and means, since both are reduced to simple conformity to the norm, to the pressure to behave "as one should", "in accordance with the rules". On the whole, however, when other guidelines are involved, it can be considered as socially accepted means. As well, the behavioural stereotype of the normative action becomes evident the way in which it is comprehended. This realization is stereotypical since it does not have an expressed rational character. Neither is the action psychologically unconscious.

Normative behaviour does not imply a social reward in each instance where one complies with the norm, but assumes that there is a reward at a level of the social system as a whole (i.e. for an individual it is contingent, not guaranteed).

Norms determine behaviour more rigidly than, say, values do. A norm is either complied with or violated, whereas one may conform to values with varying intensity [58, p. 258]. They are, however sufficiently flexible since they specify not what should be done but what is forbidden, and simply outline rather extensive boundaries of the socially permitted.†

---

*I will distinguish them from the moral norms (obligations). The most general difference between these two may be explained by the fact that "reward" and "punishment" in the first case are effected from outside (social sanctions), whereas moral norms originate inwardly ("conscience"). For more details see [11, 15].

†The norm flexibility is transparent in the well-known phenomenon of ambivalence (noticed by Merton and folklore students) when a norm is expressed in two directly opposite ways.

Norms serve to optimize behaviour (its successfulness and social acceptability) using stereotype "solutions", and to preserve the individual's intellectual, psychological and other resources.

The degree and the nature of the normativity of the above mechanisms [(1)-(4)] are different: values are basically existential and non-normative; obligations are the most inherently normative; and norms are the least existential and inherent.

(5) The goal orients a human being in the world of instrumental objects, natural and social means. The relevant (personal) meaning and the substance of the goal is wholly determined by other behavioural guidelines. The goal gives expression to that meaning through rationally-chosen external objects (processes), results, means and level of aspirations. The statement of a goal in those categories is called its "concretization", "rationalization".

Rationally choosing the objects neutral in respect to the values, norms, etc., i.e. objects which are indifferent towards the other guidelines of behaviour, is conducted in terms of their feasibility and availability. Hence the range of rational choices of goal is extremely restricted. A goal is an ideally stated (as a goal) result of the action, in other words, a motivated, comprehended and explicitly worded anticipation of a future result [51, p. 5; 53, p. 459], which is a condition and/or cause of its implementation.* It furnishes the human being with an idea of a desirable result, but not of a process or of an accomplishment (as a norm or an obligation, for example, do). The goal determines a result from the standpoint of external objective processes and phenomena (but not the internal state), i.e. from the standpoint of rationally chosen means (justified in terms of norms, values, etc.). This is why the central link in goal-seeking behaviour is not the goal but the means [3, pp. 45-46; 52]. Finally the goal explicates a desirable level of the satisfaction of needs, attainment of values, etc., i.e. a certain level of aspirations. The function of goal-seeking behaviour is in the direct and instrumental (ignoring the processes of goal-setting and employing the external objects as means) implementation of aspirations set by other guidelines.

(6) Self-oriented human behaviour is directed not to external objects, but to oneself as an individual. This existential guideline is closely related to the life project and in many respects resembles it. However, it is less generalized and integral, more concrete psychologically, and does not possess its normative nature.

Orienting behaviour to one's own individuality expresses itself in the aspiration to be (not to be), to become (not to become) somebody (or something), and to acquire or retain certain inherent properties or inner states. It is expressed in the aspiration for self-knowledge, high level of

---

*This accounts for one very specific nature of goal-seeking behaviour which should be distinguished from purposive and other types of goal-oriented behaviour. Goal-seeking behaviour is not "directed to the goal", but is "directed by the goal (from the goal to the result)" [39, p. 100]. A goal is not causafinalis but causa efficientes, not any useful result for whose attainment behaviour is effected, not a terminal situation attained in the functioning of a system, but "a real phenomenon of awareness which existed prior to the explained event and became one of the real conditions and even a cause of its emergence" [51, p. 6]. Hence the possibility and necessity of differentiating between "oriented and goal-oriented" systems (Nagel).

self-appraisal, self-actualization (self-creativity, development of one's capabilities and self-expression) and inner harmony [8]. These guideline components are mutually independent, allow for a hierarchical organization (whose nature in the final analysis determines the nature of the behaviour), may be contradictory, or may vanish or become goals in their own right. The guideline itself can also occupy various positions within the mechanism of goal-setting - either it is subordinated to other guidelines, or all of them (with the exception of the first) are subordinated to it (in this instance, self-actualization or inner harmony are considered a top priority of the value hierarchy, whereas meeting obligations or attaining goals are considered simply means of self-assertion, etc.). The main function of this behaviour type is in preservation and health of the personality.

(7) Absence of a guideline of behaviour. In this instance it is not directed to any specific object, or goal and appears to be "meaningless" or "irrational". A person cannot convincingly account for the goal of or the reason for such behaviour, does not select the means, nor anticipates any reward. External characteristics of this kind can mask pseudo-non-goal-oriented behaviour and behaviour as the search for a guideline.

In the former case behaviour has either not been rendered intelligible (by both observer and agent of the action) and rationalized (although possible), or the process of its apprehension and orientation is still in the initial development stages and is not represented at a particular moment. Then it appears impulsive or as an "outburst" arising from the instantaneous nature of the reaction. In fact, it veils a complicated system of orientation, and a prior choice, for instance, a reaction to injustice or rebellion.

In the latter instance, non-directed activity at various levels of behaviour (from physical to complex inner activity) presents a permanent departure from the former state, trespassing its boundaries, and aspiring for something different [27, 17]. It seems to be the search for an object, goal, or guideline of activity. For all its non-directedness, however, activity of this kind is never accidental or chaotic; it implies potentially oriented behaviour since its evolution is always directed outwards and not inwards. Furthermore impulsive behaviour viewed as certain needs, motives, and aspirations which have not yet been recognized or embodied as a goal also falls into the same category. Contingently it is a mere instinctive response by an individual to situations which it has neither immediate guidelines nor the possibility of outlining them at the particular moment.

Thus non-directed behaviour can be both preparation for oriented behaviour, and an involved manifestation of it. However, it may also happen to be an obsolete rudimentary stereotype of the response which either did not receive or lost orientation. The function of non-directed behaviour is a search for goals, guidelines of activity and the unconscious realization (including non-instrumental) of other mechanisms of orientation.

As I see it, the analysis of the prevailing model of goal-seeking behaviour and the description of various types of such behaviour indicates the possibility of a new approach to dealing with this problem. In this approach, the process and mechanisms of goal-setting are considered as an element of human goal-seeking behaviour. The above description of the goal-setting mechanism is only a first stage in the study of that system. The next step will be to analyse this mechanism on both the theoretical and empirical level and to disclose its systemic properties and interrelations.

REFERENCES

1. K. A. Abulkhanova, *On the Agent of Mental Activity*. Moscow, Nauka (1974), in Russian.
2. R. L. Ackoff and F. E. Emery, *On Purposeful Systems*. Chicago, Aldine (1972).
3. R. L. Ackoff, *A Concept of Corporate Planning*. New York, John Wiley (1970).
4. *Active Systems*. Moscow, IP Press (1973), in Russian.
5. M. A. Alemaskin, *Psychological Traits of Adult Delinquent Personalities*. Thesis, Moscow (1968), in Russian.
6. F. V. Bassin and V. Ye. Rozhnov, On the modern approach to the problem of unperceived mental activity (unconsciousness), *Voprosi Filosofii* 0 No. 10 (1975), in Russian.
7. G. S. Batishev, Activity substance of human beings as a philosophical principle, in: *Problem of Human Being in Modern Philosophy*. Moscow, Nauka (1969), in Russian.
8. L. Berkowitz, Social motivation: *Handbook of Social Psychology*, New York, Addison-Wesley (1969).
9. I. V. Blauberg and B. G. Yudin, *Concept of Wholeness and Its Role in Scientific Cognition*. Moscow, Znaniye (1972), in Russian.
10. I. V. Blauberg, Integrity and systems properties, in this book.
11. M. I. Bobneva, Social norms and behaviour regulation, in: *Psychological Problems of Social Regulation of Behaviour*. Moscow, Nauka (1976), in Russian.
12. H. Bossel, *Notes on Basic Needs, Priorities and Normative Change*. Karlsruhe, ISI (1975).
13. H. E. Brogden, The primary personal values measured by the Allport-Vernon test, in: *A Study of Values*, Washington (1952).
14. R. E. Catton, Exploring techniques for measuring human values, *American Sociology Review* 19, No. 1 (1954).
15. O. G. Drobnitsky, *The Concept of Morals: Historical and Critical Study*. Moscow, Nauka (1974).
16. O. G. Drobnitsky, Theoretical foundations of Kant's ethics, in: *Kant's Philosophy and the Present*. Moscow, Mysl (1974), in Russian.
17. P. Fraisse and J. Piaget, *Traité de Psychologie Experimentale*, Vol. 5. Paris, Press Universitaire de France (1965).
18. Ch. Fried, *An Anatomy of Values: Problem of Personal and Social Choice*. Cambridge, Mass., Harvard University Press (1970).
19. P. P. Gaydenko, *The Tragedy of Aesthetism: An Essay on Soren Kierkegaard's Weltanschauung*. Moscow, Iskusstvo (1970), in Russian.
20. P. P. Gaydenko, M. Heidegger, in: *Philosophical Encyclopaedia*, Vol. 5. Moscow, Sovetskaya Encyclopedia (1970), in Russian.
21. J. P. Gibbs, Norms: The problem of definition and classification. *American Journal of Sociology* LXX, No. 5 (1965).
22. L. Gorlow and Y. A. Noll, A Study of empirically derived values, *Journal Sociological Psychology* 73, No. 12 (1967).
23. D. M. Gvishiani, *Organization and Management*. Moscow: Progress (1972).
24. A. A. Ivin, *The Logic of Norms*. Moscow, Izd-vo MGU (1973), in Russian.
25. I. Kant, *Groundwork of the Metaphysics of Morals*. New York, Harper and Row (1967).
26. C. Kluckhohn, Values and value of orientations in the theory of actions, in: *Towards a General Theory of Action* (T. Parsons and E. Shils, eds.). Cambridge, Mass., Harvard University Press (1951).
27. T. A. Kuzmina, Human subjectivity as an ontological problem in modern Western philosophy, *Voprosy Filosofii* No. 9 (1977), in Russian.
28. A. Leontyev, *Activity, Conscientiousness and Personality*. Englewood Cliffs, Prentice-Hall (1978).

29. T. B. Lyubimova, The concept of value in Western sociology, in: *Social Studies*. Moscow, Nauka (1970), in Russian.
30. M. G. Makarov, *The Category of Goal in Pre-Marxist Philosophy*. Leningrad, Nauka (1974), in Russian.
31. M. K. Mamardashvili, Indispensability of form, *Voprosy Filosofii* No. 12 (1976), in Russian.
32. M. K. Mamardashvili, J.-P. Sartre, in: *Philosophical Encyclopaedia*, Vol. 4, Moscow, Sovetskaya Encyclopedia (1967), in Russian.
33. Yu. S. Martemyanov and Yu. A. Shreider, Rituals as a self-validated behaviour, in: *The Sociology of Culture*. Moscow, Mysl (1975), in Russian.
34. J. S. March and H. A. Simon, *Organizations*, New York, John Wiley (1963).
35. A. H. Maslow, *Motivation and Personality*. New York, Harper and Row (1954).
36. R. K. Merton, Social structure, anomy and deviation, in: *Social Theory and Social Structure*, Glencoe, Free Press (1957).
37. C. W. Mills, *Sociological Imagination*. Oxford, Oxford University Press (1959).
38. N. F. Naumova, The problem of goal of global systems modelling, in: *Systems Studies Yearbook 1978*. Moscow, Nauka (1978), in Russian.
39. Ye. P. Nikitin, *Explanation as a Function of Science*. Moscow, Nauka (1970), in Russian.
40. R. Norman, *Reasons for Actions: A Critique of Utilitarian Rationality*. New York, Harper and Row (1971).
41. A. P. Ogurtsov and E. G. Yudin, Activity, in: *Great Soviet Encyclopaedia*, Vol. 8, Moscow, Sovetskaya Encyclopedia (1972), in Russian.
42. V. Pareto, *The Mind and Society*. New York, Dover (1935).
43. S. L. Rubinshtein, The Person and his world, in: *Methodological and Theoretical Problems in Psychology*. Moscow, Nauka (1969), in Russian.
44. J. F. Scott, *Internalization of Norms*. Englewood Cliffs, Prentice-Hall, 1971.
45. C. Z. Sharthe, G. B. Brumback and J. Rizzo, An Approach to Dimensions of Value. *J. Psychol.*, Vol. 57, No. 1 (1964).
46. T. Shibutani, *Society and Personality*, Englewood Cliffs, Prentice-Hall (1961).
47. G. P. Shchedrovitsky, Automation of design and the problems of design activity development, in: *R and D for Automatic Design Systems*. Moscow, Stroyizdat (1975), in Russian.
48. J. Szczepanski, *Elementary Sociological Concepts*. Moscow, Progress (1969), in Russian.
49. N. J. Smelser, *Theory of Collective Behaviour*. London, Routlege and Kegan (1962).
50. V. A. Yadov (editor), *Socio-psychological Portrayal of an Engineer*. Moscow, Nauka (1977), in Russian.
51. O. K. Tikhomirov, Concept of "goal" and "goal-formation" in psychology, in: *Psychological Mechanisms of Goal-Formation*, Moscow, Nauka (1977), in Russian.
52. N. N. Trubnikov, Means, in: *Philosophical Encyclopaedia*, Vol. 5. Moscow, Sovetskaya Encyclopedia (1970), in Russian.
53. N. N. Trubnikov, Goal, op. cit.
54. V. A. Yadov, Dispositional regulation of social individuated behaviour, in: *Methodological Problems of Social Psychology*. Moscow, Nauka (1975), in Russian.
55. M. Weber, *Werk und Person*. Tübingen (1964).
56. S. V. Yemelyanov and E. L. Nappelbaum, Individual and collective purposefulness in management sciences, in: *Control and Management Problems*. Moscow, IPV Press (1975), in Russian.
57. E. G. Yudin, Philosophy versus science relation as a methodological problem, in: *Philosophy in the Modern World. Philosophy and Science*. Moscow, Nauka (1972), in Russian.

58. Yu. M. Zhukov, Values as decision determinants, in: *Psychological Problems of Social Regulation of Behaviour*. Moscow, Nauka (1976), in Russian.

Systems Approach to the Science of Science

BODY OF PUBLICATIONS AND SCIENCE DISCIPLINE SYSTEM

E. M. Mirsky

In an earlier publication [11] on methodological problems of systems research of science I emphasized the (mainly) statistical laws pertinent to a body of science publications whose substance interpretation opens way to disclosing the principal processes characterizing a science discipline as a system. The aim of this chapter is to extend the previous analysis to include the effect of structural features of the body of publications on the functioning of a scientific discipline.

In almost all theoretical descriptions of the science of research efforts, the representation of the subject matter of study is based on an empirical material derived from publications. The publication is used as a primary source of information concerning the scientific knowledge, relationships between researchers, the structure and dynamics of scientific associations, etc. Publications have been the ultimate reality for the person studying science, the philosopher, the logician, the methodologist, the information expert or (until recently) the person studying the sociology of science to derive his or her concept of a particular science. No other form of science is available to these people. The primary role of the scientific publication as a source of research is both a necessary condition and a methodological principle, even of the history of science in which documents other than the publication (unpublished manuscripts, draft versions, letters, contemporanians' memoirs, etc.) are also studied meticulously (see [25]).

The descriptions of science differing in various research traditions, i.e. new knowledge acquisition and theory-to-theory transition (the logic of the evolution of science); the concept of a paradigm, a scientific community and regulations of relations between its members (Kuhn concept); and the role distribution between those involved in research and the typology of scientists (sociology and psychology of science) are objects of study as long as information on these descriptions is available in a scientific publication.

This fact had to be mentioned since it is often not reflected in the above traditions. Discussions within the framework of each of the traditions are based on secondary material and the object is to interpret pieces of information that were earlier derived from publications of a particular science, but had been incorporated by tradition at the time of discussion. As a consequence, no question is raised practically of the relationship between the idealized objects (such as scientific knowledge, scientific theory and scientific community) of a research tradition and systemized empirics when an object of this sort is to be constructed. One postulates that the integrity of science is an intuitively given entity while any tradition studies an aspect of this integrity. This premise basically eliminates the possibility that the question of aspects interrelations might be stated in a form requiring the intuitive concept of scientific integrity to be obligatorily reinforced by a description of the science architecture.

Awareness of the fact that an empirical basis in addition to an intuitive premise is common to all the traditions mentioned is extremely significant, as this common set of empirical data can be presented in a discrete and well structured rather than in an intuitively obvious form.

The problem of explaining interrelations between various aspects of science of science research is thus stated in a more productive (in terms of research) form. On the one hand, there are independent descriptions of processes characterizing the functioning and evolution of the scientific discipline available from various research traditions. On the other hand, information is available on the architecture of a set of empirical data common to all the various traditions. The objective is then to show how this information might be employed to provide a systems-theoretical image of a scientific discipline whose varied aspects are studied by different scientific traditions.

A scientific discipline itself is considered here as a stable form of modern organization of science in all its essential manifestations (such as scientific knowledge, professional aspects of research activity, and peculiar features of reproduction and development). From this standpoint, one can deal with a scientific discipline as a relatively autonomous system with developed self-control mechanisms providing a valuable and time-stable system. Since the overall life time of the system is far greater than that of any of the system components (this property is essentially independent of whatever kind of resolution into components is used; (for details, see [11], pp. 190-193), one must take into account reproduction processes in addition to function processes to obtain an adequate description of the system.

In the discussion of the substantive interpretation of the above concepts of a discipline as a system, one needs to determine how a scientific discipline retains at any given moment its wholeness as a combination of knowledge and profession in spite of, or due to, the continuous supply of ample new research data as well as new members of the community of that discipline. In doing so, the following must be taken into account.

First, because it is a union based on a professional-and-subject ground, any scientific discipline retains its wholeness even though its living processes are distributed over space and take place in different institutional environments.

Second, although the mass of disciplinary knowledge is always being enriched by new research data, the scientific substance of a discipline at any instant can be and is stated as a compendium whose size does not prevent it being mastered by an individual to the extent enabling him or her to join in research (see [22]).

Third, the description of basic processes within a discipline must take into account the openness of the system, its permanent interaction with other disciplines, and the possibility of differentiation and/or evolution.

Fourth, while considering the organization of a discipline as a time-invariant scheme, we have in our description to take into account the (historically) temporary nature of the substance of all its components, whether relating to the composition of the disciplinary knowledge, the membership of the community of that discipline or even its criteria of being a science, the significance of estimates, etc.

This statement of the problem of the wholeness of a scientific discipline has become possible because of the vast amount of data amassed in the last 10-15 years. The scientific literature has been the object of intensive research in science of science and information science during this period (for a detailed review of basic research trends, and a bibliography see [5, 13]). These studies can with some reservations be categorized in two trends.

One trend has examined the function of the publication in scientific activity. The subject matters of the specific studies have primarily been various indicators of one type of publication: either the scientific paper or related publication events such as co-authorship, citations, publication times in various groups of journals, or interrelations of the author, editor and reviewer. These studies have demonstrated the role of the publication in motivating researchers, in supporting a concerted and continuous nature of research.

I consider it specially significant that the body of publications was found to be the most stable part of the disciplinary whole for any community member of that discipline. Nothing is deleted from this body; the disciplinary archive is unique for all the members, and is (though only in principle of course) available to any member of the discipline.

One can also attribute the stability of the scientific literature to another of its properties, i.e. to the rather stringent format requirements. The final product has always the same standard and universal form, be it an epoch-making discovery or the substantiation of a special result, a drama of ideas or a routine polemic.

This property of the publication set has been employed in an alternative trend, in which the set was processed statistically to find out significant organizational characteristics of scientific activity in general.* This means that relations within the publications set were considered as indirect indicators of the entire set of interrelations within the scientific discipline. While in the former trend, publications are the objective of research, in the latter they are only studied as an indicator of relations in a contextually wider set.

During these studies there have been great improvements in their methodological basis, powerful computer technology has been used and sophisticated techniques to interpret data developed. This has given more reliable and wider theoretically applicable results. Yet it is becoming increasingly obvious that any further progress in the scientological study of publications (and, respectively, of discipline functioning processes through discipline publication indicators) is being hampered by the inadequate structural concepts used, and the underlying object models [9, 16, 19].

---

*It should be noted that here one also dealt essentially with studying primarily the entire set of scientific papers.

Most models in the scientological study of publications are based on comprehending science almost exclusively as *research activity* and new knowledge as results of the latter. Any *increment of knowledge* such as a scientific discovery, a solution found for a problem or a falsification of theory is in one way or another a basic unit to be modelled in a specific study.

A publication is considered to be an empirically observable result of some incremental growth of knowledge and the evolution of a discipline as a time sequence of substance-related cognizance events. This sequence should represent the historical progress of the "research front". The fact that scientific efforts as well as their representation in a publication set are not limited to pure research and cannot therefore be perpetrated at the "research front" alone is generally recognized as insignificant[*] and is only accounted for in some special problems dealing with science.

A significantly more meaningful concept of the discipline structure and characteristic processes in this structure is required to implement a systems-theoretical description of a scientific discipline along the lines stated at the outset of this paper. This sort of structural unfolding is possible not only in principle or at the level of general reasoning; it can be related to systemized empirics through an appropriate decomposition of the disciplinary publication set and a functional description of set components.

ECHELONS OF THE DISCIPLINARY PUBLICATION

In the previous section it was mentioned that the above descriptions of science dealt with individual pieces of endeavour at the forefront of the discipline while the historical evolution of the discipline was again understood as the research front motion. It would seem natural to try to unfold the description of research efforts and their reflection in the disciplinary body of publications to look into them at greater depth. Without taking anything away from the terminological consistence, I will call the units to be used in the unfolding process "echelons" at different time-distances from the research front. The goal in terms of meaningfullness is to complement concepts of the discipline's historical evolution (philogenesis) by a description of what can be called ontogenetic features. The main processes in which these features manifest themselves on a full scale are the assimilation of new research data within the structure of a disciplinary knowledge, and the transfer of new members of the disciplinary community to the research front. These two processes must be described not as individual facts (this would be too great a simplification) but rather as a certain sequence of stages with an overall length of a few or possibly many years. The two processes are limited, on the one extreme, by the research front (it is here that new results appear and the "education" of scientific newcomers is completed) and on the other, by a comprehensive and systematic presentation of the discipline subject matter as seen by an external observer in academic courses, popular accounts, etc. These two processes counteract each other.

---

[*]The tendency to look for empirical references of science as close as possible to the "research front" is also reflected in the fact that as empirical research techniques have been improved in the recent years, the emphasis has been moving from the scientific paper towards its lower types of either oral or written publication (such as a report at a scientific meeting and a preprint) and informal communications.

The publication set can be decomposed along the same lines to discriminate echelons at different distances from the research front. The principal characteristic to be used in the decomposition procedure may be genre characteristics of a publication. It can be shown from the history of science that individual genres of scientific publications such as a monograph, paper, textbook and abstract appeared at different times in response to certain demands of developing scientific activity (more rapid communications between those doing research, information supply to the mass scientific profession, priority protection, etc.). (This problem is covered in [12, 21].) The appearance of any new genre did not imply that it was to replace some of the then existing ones. In fact, that new genre would complement the list of the existing genres, and the event would be accompanied by a redistribution of functions within the entire data set. From this it can be assumed that any genre fulfils specific functions different to those of other genres within the publication set (as well as within the discipline).

The idea of the functional uniqueness of a genre is based upon certain systems-theoretical considerations, and not only upon the historical features of the publications set. If one wishes to regard the discipline as a system possessing developed self-control, it must be assumed that self-control mechanisms are "visible" in some of their manifestations (such as in the form of criteria) to any member of the disciplinary community. This is where, specifically. the assumption rests that the entire set is involved in discipline control processes. The decomposition of the publications set into several function-specific components is therefore a result of (1) these components being adequately stable and (2) main functions of these components (though not the entire set of their interrelations) being intuitively obvious to all the disciplinary community.*

In order to interpret publications data subsets as echelons of a disciplinary body, I will order them according to their time distances from the discipline forefront, with the scale unit being the least amount of time required for a research front result to be published in the respective genre. The publications sequence (generalized, naturally) would be as follows: (1) papers from journals and scientific meetings; (2) data confirming earlier notes (problem-oriented, analytical, etc.), reviews of periodicals or scientific meetings for time span; (3) topical volumes, tutorial review papers, personal or collective monographs; and (4) textbooks, manuals, readers, popular scientific descriptions of the discipline. Though the substance of any of these classes could naturally be presented in fuller detail, it falls outside the scope of this paper; in fact, I will follow others doing research in the field [14, 17] in further simplifying the matter by denoting each of the classes by a single word. The resulting sequence is (1) papers, (2) review papers, (3) monographs, and (4) textbooks.

The publication echelon is an empirical equivalent (identifier) specifying a basic stage in the system of processes ensuring the wholeness of the discipline. As a consequence, the functioning of mechanisms controlling any given process is most obvious at the echelon boundaries, while the functioning of any echelon can be interpreted in terms of input/output.

---

*It is this concept of the function of each genre that is taken as a starting point in studying publications on the history of science by people from such fields as library or information science.

The term "information" is used (as applied in the information sciences) to denote the subject matter as one moves from one echelon to another. Each echelon is regarded as a common information carrier for feeding in the input, being processed, code-converted and transferred to the next echelon via the output. Genre features of a publication are considered to represent processing, organizational, and information coding techniques within each echelon.

Each echelon is simultaneously treated as a discrete set of single-type publications, with quantitative characteristics being significant in this treatment. Echelons strongly differ in terms of size, the latter decreasing in relation to the distance from the forefront, with a huge and fast growing set of papers on the one extreme, and a relatively small set of textbooks on the other. Available data [2, 17, 19] shows that in addition to the organization of each echelon, information processing also envisages a simple sorting out of publications and the amount of sort-outs varies with the disciplines, reaching up to half the echelon size in certain instances. This variation is very greatly pronounced in the data from processing papers, as a sizable fraction of papers go into archives to disappear forever from references and review papers.*

Once in the archive, a publication does not re-enter the discipline; at best, it may be discovered and re-evaluated by science historians at a later time.

While the degree of impartiality and sort-out criteria are subject matters for separate studies, the sort-out efficiency is extremely high, especially in sociological terms. It is not only the substance of the publication that is sorted out; in fact, the list of contributors and of authors referred to is increasingly reduced from one echelon to another. As the scale of information units increases, the assessment criteria applied to individual contributions change and a constantly increasing number of names (also including, under the existing citation rules, those co-authors who have bad luck with their initials) falls under the "*et al.*" item wherever the respective publications are close in their subject matter. Finally, formal references to the authors of major and well-known achievements (such as Euclid, Newton and Mendeleyev) are not a general practice in the disciplinary classics unless the respective contribution officially bears its author's name. The natural sciences differ fundamentally from the liberal arts in this respect.

Quantitative characteristics of publication echelons also greatly affect the essence of the replenishment process of the scientific community based on providing newcomers with specialization. Here, publications within each echelon must indicate the reader reference points to choose relevant publications from the data set of the next higher-level echelon.

When treating echelons of publication body or processes within a discipline, I emphasize in accordance with the goal of study, the systems nature of the subjects to be analysed. This certainly does not imply that I consider the publication set or a discipline, either as a whole or in the forms of their functioning, to be an automaton. In fact, I identify specific rules, determinants and regulations that must indicate conditions under which the functioning of a scientific community, its constituent groups or individual

---

*This parameter is also quite large at the paper echelon input where journal editors select manuscripts.

researchers develops as a reasonable effort based on making and implementing motivated (individual or group) decisions.

PUBLICATION ECHELON FORMATION AND STRUCTURE

The Paper Echelon

The input of this echelon occurs at the interface between the research results available to a narrow group of people, and the results already circulating in the standard mass media. Manuscripts coming to editors of journals describe results that have generally been obtained 1.5 or 2 years previously, have been discussed at various panels, and thought by both authors and opponents to be worth being published.

Even so, editors sort out the manuscripts received to reject a certain (possibly large) part of them. The reasons and motivations for rejecting have been subjected to a fairly comprehensive study (see [18, 23]). But far less attention has been given to what makes editors accept a paper, although this is the basic question if we want to know how the paper echelon is formed.

I will begin with a few general comments. Most editorial boards consist of leading experts in the field who vigorously strive to make their respective journals instrumental in the further development of the discipline by providing the scientific community with timely and very detailed information on every research endeavour of potential interest. When asked about the motives for rejecting manuscripts, editors least of all indicate the lack of space. In fact, they are inclined to think that even among those selected, too many are of no significance and are fillers rather than information carriers [14]. To confirm this, reference is made to information indicating that much of the published material evokes no response, and therefore does not participate in the information exchange, going instead right to the archive of science. The following items are most often listed among the reasons for rejection of a manuscript: the results presented are trivial; the conclusions and interpretations are ill-grounded; the author's knowledge of the work in the field is inadequate; the content of the manuscript falls outside the journal's scope. I will consider each of these arguments in detail.

The triviality (or originality) of a result appears to be the most obvious and impartial feature. Triviality means that the result in question has been published earlier and can be presented to the author as a reason for rejecting his contribution. However, things are not as simple in most instances, for there are very few manuscripts fully duplicating results known earlier.* More often it is asserted that the result presented is not original enough which implies that the corresponding information is insignifcant. However, what is involved here is not an unbiased criterion but rather the reviewer's opinion supported by his knowledge and his understanding of what the most promising ways of studying a particular problem are. This judgement clearly diverges from that of the author, and of those who took part in previous discussions. In fact, it often clashes with the opinion of editors of another journal to which the author sends the manuscript once rejected by the former [14].

---

*Admittedly, these manuscripts may be encountered and published, and this fact in the history of science has several times been a source of litigation over priority.

The following point should be underlined in this context. The editor or reviewers (who, as was mentioned, have the highest expertise in the field) have to solve quite a simple (in the opinion of many sociologists of science) problem – to discriminate, in their own field, between fresh knowledge and previously existing knowledge. However it turns out that knowledge exists at the research front in such a form and changes so rapidly that it is impossible to be objective regarding the criterion of novelty. Knowing what the available knowledge is, is not enough to decide whether a specific actual result is original or otherwise (this is the subject matter of the paper in question), since any practical assessment process takes into account a potential future development of that knowledge. This is why absolutely inevitable errors result in (1) conflict situations and (2) papers published that are not to be subsequently used by the scientific community.

The situation is similar for other criteria used in selecting material to form the paper echelon. Thus, the rejection of a paper on the grounds that its conclusions are not substantiated, or that the author's knowledge of other works in the field is inadequate, might also be based on the editor's (or reviewer's) decision related to the author's interpretation of other works.*

It is this decision that results in the final judgement over the numerous "evidentlys", "hopefullys" and "most likelys" that abound in any manuscript. If one bears in mind that we are discussing interpretations, another conclusion becomes evident: that the requirement of conclusion validity largely contradicts the concept of novelty, since any well-substantiated specific result, except for purely empirical results, is very likely trivial.

This becomes obvious from the research data provided by Whitley [18], who studied relationship between the type of paper and its chances of being rejected by journal editors. Those least often rejected were found to be papers presenting some empirical data obtained through an established procedure and based on a simple and well-presented working conjecture. This type is followed by technical papers. Most often rejected are theoretical papers with a detailed interpretation of data.

It was mentioned previously that a widespread explanation of why a manuscript was rejected is that the manuscript content does not dovetail with the journal's area of interest. In other words, it is said that in sending in his paper the author was not simply misled about some minor aspects of the editors' position, but rather failed to determine the basic coverage (the area of interest) of the particular periodical. The inevitable conclusion then is that the area to be covered by future issues of the same journal cannot be predicted with adequate certainty from what was published earlier. This in turn implies that (1) the journal's area of interest is changing rather rapidly and (2) the editors' idea of what imminent changes might be is based on future trends of the profession rather than on its history.

---

*One significant aspect is to be noted. Insistence on a meaningful discussion with the manuscript author about his colleagues' works implies that the editor demands that the author clarify where he thinks the novelty (or the potential) and validity of the new contribution in his narrow area lie.

By this I do not mean an intuitive prediction. The editors (unlike the manuscript authors) have fairly complete and reliable data for an idea of what their journal might become in the future. Some of this data can be found in the journal's portfolio, that is, in the manuscripts to be published. As well, editors and members of advisory boards are very highly qualified experts in an area of research, and are fairly well aware of what work is underway at the research front and will only be presented in the journal a few years hence (the time required to complete a research project and to process the results obtained for publication).

It was necessary to present so much detail about the input characteristics of the paper echelon since it is the only echelon whose input materials are selected through communications between experts and at which level the decisions taken must somehow be explained (and therefore demand some reflection). A result which is currently in the course of being assessed cannot be compared with a canon common to the entire community (true knowledge) neither can rigorous and unbiased criteria (such as validity, uniqueness and promising features) be applied to the result.

This is why that by the time of its appearance in print the paper as such* is just a piece of correct information about a research result instead of being a quantum of new knowledge. The correctness requirement applies to the technique by which the result was obtained and published and to the result itself. There is no room for revelation in a scientific paper. A scientific paper simply informs that the new (in the author's opinion) result was obtained in conformity with the existing standards of research and is published in conformity with the code of ethics adopted in the particular scientific community (see [23]). This can clearly be seen from the paper "anatomy".

"A journal paper", say Nalimov and Malchenko, "consists of the following parts: a list of authors, the author's(s') affiliation and sources of support; the title claiming to represent the content in the shortest possible form; an abstract or summary; the paper itself; acknowledgements and references" ([13], p. 96).†

At least four of the above items are purely sociological, as they indicate the author's position within the scientific community and his relationships with others doing research and/or their groups, while describing no substantial property of the result in question. Any paper tells what research is underway at the forefront at a given time and what results have been obtained.

The paper structure and format are designed so that the reader most readily obtains information along any one of these lines. If he is interested in the result as such, then the paper title is at hand as well as an abstract

---

*In this context, the actual status of a paper in the publication echelon differs as much from the status of the same paper analysed by the history of science many years later as a just purchased lottery ticket does from a winning one, though the same paper is apparently implied.

†Many journals also indicate the name of a well-known scientist (though not an official reviewer) who recommended that the specific paper be published and is therefore willing to share with the author and editors responsibility for the validity of the paper's content.

(since any author is obliged to summarize only new data in the abstract) and the very appearance of the paper in a journal is a validation of the result. If, on the other hand, the reader is primarily interested in the research state-of-the-art, the paper tells him who, where, in what team, by what techniques, and using what references, has carried out the particular research. This entire information chain must be constructed for a relatively short and scope-limited contribution.

It is necessary to supply every paper with this wealth of information indicators, since the paper echelon as a whole has a very low level of organization and, in fact, very limited opportunities in this respect. The entire set of journal papers introduces a set of independent generalizations on the state-of-the-art at the research front in publications. The paper echelon reflects the state-of-the-art within the discipline. On the other hand, attempts to structure the echelon content into levels or blocks have so far been hampered by enormous difficulties. I touched on the nature of these difficulties when discussing aspects of publication echelon formation.

Thus, the most standarda and obvious type of publication echelon decomposition problem is to classify this echelon, within a discipline, into topical journals, each with its predominant scope. However, we saw that no definite or content-stable area of interests of a journal can adequately be outlined - not only for the readership, but even for the journal prospective authors, and these often send their contributions consecutively to various journals within the discipline. The content of journsls often overlap, thereby producing what is called interfacial areas. It has already been noted that this primarily results from the fact that any journal publication represents the discipline forefront dynamics, and this dynamics is determined by the advances in research rather than by the scope of the journal. Any change in the forefront also affects its structure, thereby requiring a respective change in the entire paper echelon structure even if this structure had no similarity in the earlier journals differentiation with respect to subjects covered.

The situation is equally difficult if the scale of topical differentiation is reduced to the level of topical sections within journals. More specialized, and therefore more closely related to the existing topical differentiation of problems, the topical-sectional components of the structure are even less stable, since lower-scale problems vary faster than any profession represented by a journal. Unstable journal sections bewilder the reader still further as he must keep close tabs on not so much the papers, as the varying journal sectional structure. This is why many journals on the natural sciences have in principle abandoned (at least topical) sectionizing and only retain sections in concluding (say annual) issues intended for archives.

What is of significance here, however, is not dynamics. The nature of the difficulties when faced by attempts to structurally decompose the paper echelon is primarily that the organization of any paper has components of two different "sets", one being the topical differentiation of problems within the discipline and the other being the actual research efforts within the same discipline. Their different rates of change further heighten the problem, since they make any method intended to remedy the situation "on the spot" inefficient. This is why any paper is transferred to the paper echelon "output" with a set of information indicators which make it possible to group papers into any one of the two above lines.

Real areas for this kind of grouping are (1) the discipline archive and (2) the review echelon following the paper echelon. The discipline archive is a complete set of all published material whatever the genre, publication date, etc. The archive input is assumed to be various services supplying information or abstracts. Topical material differentiation common to any publication type has been developed and is present in the archive, along with purely functional classification techniques. Topical differentiation is based on the structure of the discipline substance* and supported by fairly stable classifiers. However, this stability encounters enormous problems in information processing and requires that many new information indicators be sought after to be used in coding and retrieval (compiling thesauruses, etc.). As a result, more and more cumbersome information systems equipped with the most sophisticated computer technology - although not intended for an "unarmed" user - are required to provide archive efficient functioning. In this respect, the archive opposes the echelonized set of publications actually functioning in the framework of the discipline.

The Review Article Echelon

This echelon is composed of information units built of paper content. These building blocks are thus a result of selecting and organizing the paper content. This work is done by experts, whose authority, past experience, knowledge and intuition is expected to fill in the obvious gap in the information characteristics at the echelon input.† The expert in the profession, in his role of a review article author has a wider spectrum of objective data available to him than does a manuscript reviewer (these roles are often combined). The review article echelon input is at a greater time distance (two or three years) from the discipline forefront than the paper echelon input. This enables the author of a review article to feel more certain about the future prospects of the results published (confirmation of experimental data or testing a hypothesis in a later study), as well as of the early response to the paper reviewed, from the communities (references).

Far from all the papers published during the time interval reviewed are mentioned in review articles. This results in a sizable reduction of the papers set. A reference in a review article indicates that even given all the years elapsed after publication, the substance of the paper referred to has an actual (and not only archivous) value, and has not been struck off by subsequent publications.

Any review aricle, as well as the entire review article echelon, reproduces the dual nature of the substance mentioned above in the section on papers. The substance-organizing process is entirely different in writing a review article. The structure of a review article reflects two lines: (1) major problems are stated in accordance with the conceptual decomposition accepted for the discipline and (2) the guideline thus formed is used to group the

---

*The process of shaping these ideas will be considered in our subsequent discussion of the monograph/textbook echelon.

†I consider here echelons of disciplinary publications. However, in major sciences such as physics, chemistry and biology, problem-oriented review articles play the role of the main *up-to-date* publication. Much space is given to various kinds of review articles in such journals as *Uspekhi* (*Advances*) *in the chemical, biological, ... sciences.*

information on different approaches and results. The latter guideline has an absolute priority, since whether and in how much detail a problem is covered depends on whether data on the respective research effort and the progress made are present in the paper echelon and not on the problem's theoretical significance (its place in the context of disciplinary knowledge).*

Organized in review-article echelon units are also the data on researchers' names and groupings, and therefore on the directions taken by the research communities. This data cannot basically be represented in the paper echelon [4].

Like the paper echelon, the review article echelon as a whole is a momentary photograph of the research efforts within a discipline (with a delay of five to seven years). It lists the directions of the most active and fruitful efforts. The state-of-the-art of each direction is represented by its latest review article, while the total echelon depth is fairly small (about three years).

An important organizational quality of the review article echelon is the fact that the papers compiled are centred around problems whose list (this might not be fully implemented) and statements conform to the basic ideas of the subject matter of the discipline.

Thus, the review article echelon (1) in a concise and operatively usable form comprises a certain number of most fruitful paper publications and (2) performs the basic organizational functions by relating the information on trends in research efforts to a conceptually stated list of disciplinary problems. In this context, the review article echelon plays the role of a link between a collection of reports on pieces of the research front (the paper echelon), and a theoretical description of individual disciplinary problems (the monograph echelon).

The Monograph Echelon

The size of this echelon is far smaller than that of either of the two previously mentioned echelons. The number of monographs (the actual value varies from one discipline to another) is 20-50 times fewer than the number of papers [16]. It thus appears that selection and reduction (i.e. a more compact organization) of the material available again takes place during the formation of this echelon. The principles and techniques used in this process are entirely different from those in the two preceding echelons.

The echelon unit - a monograph - is a systemized treatment of a basic topical problem of a discipline. The problem statement and presentation, as well as the extent to which new information from the papers and/or review article echelons is employed in a monograph, determine specific features of the monograph echelon and depend primarily on such factors as the theoretical state of disciplinary knowledge and the conceptual significance of a particular group of items and not on how intensively the problem involved is being studied at the forefront.

---

*The fact that the assessments in a review are "material-related" can be seen clearly in the manner of writing a review article. No viewpoint is disputed; rather, all the viewpoints are ranged according to the "respectability" of the corresponding group of articles.

The availability, in the two preceding echelons, of information about the problem is also very important, since the amount of new data determines what problem will be chosen for further study. However, different (i.e. theoretical and methodological) criteria of information choice and processing play an active role in the study itself. The work on new information is primarily a critical analysis of the substance of the new data (its validity, theoretical power, interpretability, etc.), while it is essentially of no significance by whom, where, when, and on what occasion this data was obtained.*

The content of a monograph is a generalization of results concerning a major problem. It always assumes a stratified theoretical and methodological analysis. I will leave aside the methodological substance and techniques used in the analysis (as these questions are covered in a very great amount of special literature) and confine their discussion only to the significance of that substance for shaping the disciplinary knowledge. To be able to treat his problem in a generalized and systematic way the author must localize it within a wider disciplinary wholeness and, therefore, outline at least the latter, with due consideration of the new understanding of the problem. On the other hand, an external standpoint is required to analyse the theoretical substance of the problem. This standpoint should be more general in respect to the object to be analysed and should be specified in terms of analytic tools, hence in the domain of scientific methodology.

The direction to be followed in processing a particular information type depends on whether a hierarchy of concepts, such as a (possibly implicit) prerequisite of analysis, is available. What is meant here is the search for opportunities to furnish an adequate theoretical interpretation of empirical data, a disciplinary methodological interpretation of theoretical issues, and a general scientific, or even philosophical interpretation, of disciplinary methodological issues. Only this systematic treatment of the entire set of theoretical and methodological problems seems to be required to introduce the information published in papers and initially processed in review articles into the body of the disciplinary knowledge. This introduction is accompanied by the re-structuring of some ties earlier established within the disciplinary knowledge.

Hence, a complete list of theoretical problems of the discipline, with a systematic discussion of the methodological substance of each problem, is represented in the monograph echelon as a whole. In fact, each unit of this echelon contains an outline of the subject matter of the discipline and describes what position the problem treated in the monograph occupies therein. Nevertheless, it is usually impossible to construct (as a rigorously stated theoretical whole) the subject matter of the discipline from its patterns available in the monograph set. This situation has been subject to logical and methodological analysis in many studies and I would like to make some comments on the aspect of this situation as a science

---

*It is not theoretism that provides the difference between a monograph and a paper, since part of a paper also discusses the theoretical aspect of the problem involved. Yet, this theoretical analysis is aimed at the problem itself (that is, at the problem statement format, new problem tackling techniques, etc.) and does not usually touch the possible revision of the problem position in the discipline context, since this methodological issue is far afield of the problem as such. On the other hand, both kinds of work are typical of a monograph.

related to certain peculiarities of the disciplinary substance existence within the publication set and its echelons.

It was mentioned previously that the total set of disciplinary problems is represented in the monograph echelon by a list of units of this echelon, with each problem represented by the latest monograph covering it.* With this list and the publication date of each monograph available, one can obtain a quantity describing the echelon time depth. This quantity is the first and the last monograph, respectively, in the list; this means that it represents the time delay between patterns of individual disciplinary problems. The time delay is caused by the nonuniformity of the above mentioned research at the discipline forefront and therefore by the nonuniform supply of information about individual problems to the paper and review article echelons. This means that later monographs give systematic analysis of a problem involved, taking into account new data from the forefront, and, what is equally important, using new analytical tools developed during the time interval between publications of two successive monographs.

However, in addition to theoretical treatment of a problem, any monograph includes a regular analysis of logical and methodological reasoning in favour of that treatment (the position of the problem in the disciplinary knowledge system). The time nonuniformity of the list of problems then turns into the heterogeneity of its substantiations. The substance of monographs written during a ten-year period on various problems of a discipline involves various (alternative) understandings of the discipline subject matter in general, as well as various methodological approaches to its study. The task of constructing a discipline from the list of its problems represented in the monographs set cannot therefore be reduced to theoretical comparisons of these problems on a common methodological basis, nor to the construction of a common logical language.

Strictly speaking, the above problem cannot be solved even if one moves back in time, for example, by constructing the subject matter from 30-year-old concepts about all the problems in the list. Though a complete image of the discipline subject matter can be constructed in this case, it will be of no theoretical significance, since it will not meet modern theoretical and methodological requirements. Efforts along these lines are therefore made with other objectives in mind. With all the theoretical and methodological expenses taken up by these efforts, a complete image and a balanced presentation of the discipline subject matter is absolutely necessary for a tutorial description of the subject matter of the discipline.

The Textbook Echelon

This echelon of disciplinary publications is primarily unique, in that the destination point of its substance does not lie inside the discipline, whether the latter is understood as a scientific community, or as a set of research efforts undertaken by community members. In fact, the idea of a discipline as presented in a textbook is intended for an observer outside the discipline who is not necessarily expected to enter into some sort of relationship with the disciplinary research in the future.

---

*I will assume roughly that the latest monograph theoretically supersedes the history of the study of the problem under consideration.

This pragmatic approach drastically affects the techniques and principles underlying the textbook design. A textbook presents the substance of the discipline in a systematic way* conforming to the general background and future profession of its addressee. The latter aspects determine the size and character - but not the content of the textbook. Thus, the main goal of a textbook is to introduce a general idea of the discipline, and of its specific features.

These qualities of a textbook are especially prominent in manuals and courses intended for college students. A discipline appears to newcomers as a whole in many respects, whether its subject matter, specific research aspects of the profession, list of luminaries, or history of a milestone.† As a consequence, priority is given to the discussion of the position of the discipline among other sciences [8, 3].

This combination of problems requires that disciplinary knowledge should have an organizational level that cannot be attained in any echelon (including the monograph echelon) representing the actual state-of-the-art of the discipline. Thus a whole description of the discipline substance, on the one hand, is systemized purely empirically, while on the other hand, statements of certain problems at the time of writing a textbook lag significantly behind their research-oriented statements in the monographs echelon. The results published by the Austrian Gehmacher are indicative. He analysed the content of general-type textbooks (textbooks on high school physics), and provides the following time lags between discoveries of major laws of physics in the nineteenth century, and their respective inclusion in the textbook:

The Ampere laws (discovered in 1820) were introduced in the textbooks in 1859 (a 39-year lag);
the Doppler effect (discovered in 1842) was introduced in the textbooks in 1927 (an 85-year lag);
the law of conservation of energy in nonmechanical phenomena discovered by Meier in 1845 was introduced in the textbooks in 1903 (a 58-year lag);
the electromagnetic light theory discovered in 1871 was introduced in the textbooks in 1903 (a 32-year lag) [20].

A noteworthy point is the nonuniformity of lags, not the lags as such. This means that different components appear in the "tutorial" wholeness of the discipline subject matter in defiance of both the up-to-date understanding of the theoretical development of the discipline and their respective sequence of appearance in the discipline subject matter.

---

*If, in a discussion of modern principles of designing tutorial literature, one points out that any textbook presents ready-to-use knowledge (e.g. [15]) this by no means implies that the problems considered in a textbook have thoroughly been studied, but rather indicates that it is possible that their complete image might be included in the kind of wholeness of disciplinary knowledge which lends itself to further euristic processing.

†This aspect was noted by Kuhn, when he spoke about studying paradigms as a way to prepare a student to enter membership, while in his definition of the paradigm itself, he does not confine himself to its purely theoretical components [6].

FUNCTIONS OF THE PUBLICATION SET

When I stated the problem in the introductory section I defined a scientific discipline as a stable form of scientific organization in modern times. The set of publications was chosen as a space wherein standards and regulations governing systems behaviour are concentrated. In the discussions of the formation and organization of disciplinary publication echelons, I attempted to characterize each of the echelons in terms of the standards and regulations that are self-evident for the community and whose implementation underlines the activity of people contributing to the formation of the echelons. Now I would like to use the above analysis to systemize the concepts concerning the way the organization of the publications set helps fulfil functions of this set in the scientific discipline.

It appears appropriate to begin by pointing out that joint efforts to form an echelonized publications set make it possible to isolate a relatively small and basically tractable group of publications from the entire mass of the disciplinary archive. This group will only involve relatively new publications, namely those whose substance has not been introduced in the subsequent echelons through selection and processing.* This group functions in a real way as a part of the publications set at any given instant. Therefore, the actual set of units in each echelon, as well as in the entire publications set (i.e. the list of publication titles), permanently varies. In other words, what is meant is an information flow with filters and converters at various stages in the form of efforts by scientists contributing to the formation of the echelons.

The decision on possible selection of a publication for further information processing (i.e. for storing certain substantial components in the publication set) are determined by certain criteria. The flow dynamics is based on the fact that the criteria of information selection used to shape an echelon and the information assessment criteria within the echelon (the latter also serve as selection criteria to shape the next-level echelon) do not coincide and, in fact, are in a sense, mutually contradictory. The substance of a manuscript received by the editor of a journal is first assessed using the correctness criterion (otherwise, the manuscript will not enter the papers set). In contrast, the substance of a paper is assessed on the basis of the fruitfulness criterion (otherwise, this paper will not be referred to, and therefore will not enter the review articles set). Units for the review articles set are selected using the fruitfulness criterion; however, whether they do or do not enter the monograph set depends on their respective validity, etc.† It should also be mentioned that the specific substance of any criterion varies as the discipline develops. The reasonableness of decisions as judged by the community is supported by the expertise and reputation of the people (such as journal editors and authors of review articles or monographs) making the choice.

The generality and structure of the disciplinary publications set are of great significance in consolidating and stratifying the disciplinary community. If a particular name from this community appears in several publications echelons, he has been recognized and his contribution has met

---

*The time depths of the echelons and of the entire set were discussed above.

†For a theoretical interpretation of the contradictory nature of assessment criteria as a general property of any generating system, using the canonical art as an example, see Lotman [7].

with appraisal. The appraisal process follows two guidelines. One assesses the research result as a contribution to the development of the disciplinary knowledge. This assessment is effected through citations in later publications [9]. Publications in different echelons are far from equivalent in these terms. For example, a single reference in a textbook is worth, for the community, a few dozens of references in journals. The second guideline relates to the great respect paid to the direct involvement of a community member in the formation of some of the publication echelons, to his or her activity on an editorial board or as an author of books, etc.* To digress from specific features of each of these status-cumulating guidelines, I would like to emphasize that implementation of each of the two guidelines is only possible because of the existence of an echelonized publications set common to the entire discipline.

Thus, the content of the publications set provides the most immediate idea of the actual state of the discipline as a whole, that is, the idea of items such as the level attained of the comprehensive description of the discipline and its tutorial aspects (the textbook echelon); the state of regular treatments of major problems (the monograph echelon); the strongest problem-approach trends (the review article echelon); and the disciplinary community stratification and forms of cumulating the status.

This information is important in drawing new membership for the discipline community from among younger scientists, as well as stemming from intra- and interdisciplinary migrations of nature research workers. The way units are organized within the respective echelon enables a migrant to advance at the greatest possible rate to the scientific forefront, while confining him to familiarization with increasingly narrowing (in terms of content) and specialized information units. The number of stages required depends on the migrant's initial background. A newcomer must go through the entire sequence, starting from studying textbooks.† For a specialist wishing to change the direction of his or her research while remaining within the same area, it is enough to learn the content of a set of articles or a review article. In this respect, an attempt by Hagstrom appears interesting in interpreting the "distance" between professions and disciplines in terms of time required by a person to begin high-level work in a different discipline [21]. Unfortunately, that attempt has not become a topic of regular research.

The rigid organization and the intensifying activity characteristic of the publications set and relating to the necessity of implementing the numerous and various functions arise due to the fact that the disciplinary body is situated within a relatively short section of two counterprocesses that develop, largely independent of each other, outside the discipline or even outside science. What is meant here is, on the one hand, the translation of generalized forms of human experience worked out in the discipline (i.e. knowledge of laws of reality and of objectivized samples of activity) to the system of culture (other disciplines, other areas of professional activity, the educational system, and the long-term social memory), and on the other

---

*The two guidelines are explicit in criteria used when nominating candidates for honorary positions in disciplinary professional associations or personal communities, such as an academia.

†A newcomer may not necessarily be a graduating student. Prominent figures such as Ashby and Delbrück had to start their activity in a new discipline by reading textbooks.

hand, the process of reasonable professional employment, in the discipline, of persons chosen by the society. The social efficiency of the two processes (or a single process, if viewed from a social standpoint) appears to ensure stability of both the disciplinary organization and the discipline as a system.

One might say with certain reservation that the former process imposes certain external limitations, while the latter does entail certain internal limitations on discipline functioning. The former process is efficient in that it develops and replenishes rapidly, and with as small loss as possible, a complete structured and objectivized output image of the disciplinary knowledge through the inclusion of research data obtained at the forefront, and presented in an appropriate form. This discipline product best suits further processing translation into an academic course (technical, medical or agricultural, etc.), handbook, encyclopedia, etc. The data that failed to appear at the discipline output (those channelled into the archives by the selection procedure) essentially stop functioning whatever their scientific potential.

The goal of our discussion of human activity within a publication echelon was to demonstrate that the tremendous efforts in selecting and/or organizing each of the echelons, including the entire process of shaping new knowledge (i.e. what appears as knowledge *both* inside and outside the discipline), take place *at a distance* from the research front [2]. Only top experts in the field participate in these efforts, unlike research efforts proper, which are largely undertaken by junior research workers and assistant personnel.

Of course, we are least inclined to under-rate research efforts proper for neither discipline development nor publication mechanism functioning could be possible if they did not exist. The point is that both the historical emergence of research activity as a separate profession and the forms of its present existence are largely determined by the disciplinary structure of science and by the division of labour imposed by this disciplinary structure. I mentioned earlier the role of publication echelons in controlling the researcher specialization process and providing the possibility of rechannelling within a discipline. The publications set occupies only a section within this process. Strictly speaking, involvement in research efforts is achieved through other types of communications not as isolated from the research front. The publications echelons only guide a migrant by telling him, apart from the characteristics of the discipline substance, who has recently done what. This is simply another essential feature of the publications set, namely, that the research result substance is far less stable (and hence provides grounds for less likely predictions) than the research worker's interests at the transition between the papers echelon and the forefront. The sociological information in the paper is then not only necessary, but also provides the most reliable road signs, if one moves towards the forefront [5].

It should be noted that no type of creative activity has until now been found with a form of information on the efforts made and results achieved equivalent to that of the disciplinary publications set in terms of efficiency in relation to both the entire community and each community member. The rapid and efficient functioning of this set is particularly obvious if compared to the giant efforts and expenditures required to improve the information performance of the current research and technological activity, and less than moderate results obtained.

Finally, a few methodological comments on the study of science and scientific discipline, and how they are mapped into concepts of science. I noted in the introductory paragraphs that these concepts are primarily based on the scientific publication as a primary source of empirical data on the object studied. The above analysis suggests a more exact statement based on the facts which emerged in the analysis, such as the specific typological and functional features of publications within each echelon, the absence of the "publication in general", and as a consequence, the distributive nature of scientific efforts aimed at forming a disciplinary whole.

These specific features apply primarily to object abstractions used in studies of science development. Thus if the notion of knowledge in its logical and structural definiteness is used as an object abstraction, the content of the monograph and textbook echelons can be a source of empirical data. However, empirically substantiated in this way can only be problems relating to the science knowledge structure and the validation of this structure, or problems of the disciplinary knowledge involvement in a wider scientific and cultural context. As for the complete set of problems relating to *the acquisition of scientific knowledge as a research function* (that is, to the scientific creative process and its "logic", a scientific discovery, etc.), the results of these forms of activity are (1) represented in other types of publications and (2) obtained and assessed in the domain in which other criteria are valid.

Things become further confused if discrimination between the available and new knowledge is introduced. Either knowledge having theoretical coherence (purely factual fragments are implied) is rejected and the problem of determining novelty in actual scientific practice therefore becomes uncertain. Or new knowledge takes the form of a totally restructured theoretical system rather than an individual fragment. *New knowledge* is then opposed not to the available knowledge, but rather to the *previous knowledge* that existed before the re-structuring. New knowledge is, however, a result rather than the cause of development in both these cases.

The specific nature of organization of knowledge in certain publications echelons (textbooks and monographs) leads to specific difficulties if one wishes to describe the history of the discipline. On the one hand, there are case studies in individual disciplinary problems, with every stage of the study of the history of a particular problem dated from appropriate sources, while essentially no question of relating the problem to a general disciplinary context is raised. The goal is to reconstruct, on a well-documented basis, the transition between problem states. The matter is essentially different if the history of the disciplinary knowledge as a whole is to be described. An interrelated description is then required for the complete set of problems at each historical stage (this will be theoretical description of the disciplinary subject matter). On the other hand, the periodization in the history of the discipline as a whole should coincide with the respective periods in the studies of individual disciplinary problems. The adoption of external indicators as the basis of periodization (that is, a periodization by centuries, epochs or world patterns) is nothing more than palliatives which do not eliminate problems. It appears that an analysis of historical aspects of the evolution of the disciplinary knowledge as carried out in the textbooks and monographs echelons, even though this does not eliminate the difficulties, might assist in understanding their nature.

It can clearly be seen from the above discussion that similar difficulties usually arise if one attempts to take the concept of human activity as a basic abstraction. These difficulties are further heightened by the fact that a related concept is usually that of human activity, which is not present in the publications empirics in any complete form. Even though the use of additional sources (such as memoirs and biographies) help elucidate some points, they cannot substitute for a regular picture. As for the reconstruction of human activity from publications, this has been seriously invalidated by recent research (see [5]). Comparison of results of a field study of research workers' activity and the picture of this activity generated in publications several years later demonstrates that many substantial and methodological problems must be solved before a satisfactory reconstruction can be achieved. Even in that case, data on research efforts would provide only a partial picture, whose understanding would be determined by the level of understanding of all types of research activity within the discipline system.

Summarizing, the understanding of the discipline as a form of science organization supported by a regular description of the publication empirics makes it possible to relate individual concepts of the discipline to the phenomena interpreted in these concepts. It is seen that what in fact is implied is not aspect views of an object, but rather the study of separate parts which cannot be interrelated until a comprehensive picture of the scientific discipline is available. I believe that this fact is significant, in that it helps to elucidate basic problems of shaping the subject matter of the science of knowledge [10] and use the substance of special concepts of science in research on the science of science. The establishment of this interrelationship might contribute to the development of the particular special concepts. Yet it is for the people representing these concepts to judge on this point.

REFERENCES

1. Z. B. Barinova, R. F. Vasil'ev, *et al.*, The study of scientific journals as communication channels: Evaluating the contribution of different countries to world progress, *Nauchno-Tekhnicheskaya Informatsiya* Ser. 2, No. 12. (1967), in Russian.
2. H. V. Wyatt, When does information become knowledge?, *Nature*, 235, No. 5333 (1972).
3. V. I. Gor'kova and A. I. Mshvelidze, A procedure to determine relationships within a discipline in an academic course, *Nauchno-Tekhnicheskaya Informatsiya* Ser. 1, No. 2 (1974), in Russian.
4. E. Sh. Zhuravel' and G. V. Korsunskaya, Classification of review-articles, *Nauchno-Tekhnicheskaya Informatsiya* Ser. 1, No. 7 (1974), in Russian.
5. *Communications in Modern Science*. Moscow (1976), in Russian.
6. T. Kuhn, *The Structure of Scientific Revolutions*. Chicago (1962).
7. Yu. M. Lotman, The canonic art as an information paradox, in: *The problem of the canon in ancient and medieval art of Asia and Africa*. Moscow (1973), in Russian.
8. A. A. Lyapunov, The educational system and a science systematization. *Voprosy Filosofii* No. 8 (1968), in Russian with English summary.
9. V. A. Markusova, Comparison of the citation index of scientific and technical publications, *Nauchno-Tekhnicheskaya Informatsiya* Ser. 1, No. 1 (1973), in Russian.
10. S. R. Mikulinskii and N. I. Rodnyi, The position of the science of knowledge within the system of sciences, *Voprosy Filosofii* No. 6 (1968), in Russian with English summary.

11. E. M. Mirsky, A systems approach to the study of science: Methodological comments, in: *Systems Research Yearbook 1973*. Moscow (1973), in Russian.
12. E. M. Mirsky, *The Information Situation in Modern Education*. Kiev (1970), in Russian.
13. V. V. Nalimov and Z. M. Mul'chenko, *Scientometrics: A study of science development as an information process*. Moscow (1969), in Russian.
14. Nan Lin, W. D. Garvey and C. E. Nelson, A study of the communication structure of science, in: *Communication Among Scientists and Engineers*. Lexington, Mass. (1970).
15. Science and the curriculum, *Sovetskaya Pedagogika* No. 7 (1965), in Russian.
16. T. M. Petrova, Methodological aspects of identification of science building blocks, in: *Systems Research Yearbook 1975*. Moscow (1976), in Russian.
17. D. J. de Solla Price, *Little Science, Big Science*. New York (1963).
18. R. D. Whitley, The operation of scientific journals: two case studies in British social science. *The Sociological Review* $\underline{18}$, No. 2 (1970).
19. Yu. A. Shreider and M. A. Osipova, Some dynamic models in information science, *Nauchno-Tekhnicheskaya Informatsia* Ser. 2, No. 8 (1969), in Russian.
20. E. Gehmacher, *Wettlauf mit der Katastrophe: Europäische Schulsysteme*. Wien (1965).
21. W. Hagstrom, *The Scientific Community*. New York (1965).
22. H. W. Menard, *Science: Growth and Change*. Cambridge, Mass. (1971).
23. R. Merton and H. Zuckerman, Patterns of evaluation in science: institutionalization, structure, and function of referee system. *Minerva* $\underline{9}$, No. 1 (1971).
24. G. Radnitzky, *Contemporary Schools of Metascience*. Göteborg (1968).
25. R. Taton, *Histoire général des sciences*. Paris (1963).

THE HOLISTIC PRINCIPLE IN AN INTERDISCIPLINARY STUDY OF SCIENTIFIC ACTIVITY

A. A. Ignatiev

Primary data treatment is one of the major problems arising in a study of research. Social sciences have now accumulated an impressive range of observational procedures, sufficiently productive and reliable to meet reasonably stringent requirements. Therefore, empirical data, characterizing the direction or parameters of certain processes may usually be obtained by using standard methods. Quite different is the state-of-the-art as regards the use of empirical data to make or justify substantive statements relative to the phenomena being observed. As far back as 1950, at the sixth International Congress on the History of Science de Solla Price first reported his measurements leaving no doubt as to the reproducibility of his growth curves of the scientific literature corpus. Yet the interpretation of these curves as parameters characterizing knowledge generation processes remains debatable to this day.

Substantive theoretical conceptions, developed in some specialized field of research, as for instance in social psychology or informatics, may be used in principle for primary data assessment. Such a strategy, usually referred to as reductionism, may pay off and even be effective as long as the study is essentially empirical and descriptive, i.e. intended primarily for recording the phenomena under observation. But it proves completely inadequate when the intention is to proceed from statements about individual phenomena, remarkable in one respect or another, to statements about the totality of phenomena under observation (particularly if such statements are to become normative in the course of time). Indeed, data confined to just one arbitrarily selected aspect of observation tend to be one-sided and therefore less than fully adequate in terms of their relevance, representativeness or comparability with other categories of data. This, in turn, not only restricts the possibility of their use (e.g. for the development of an optimal management strategy) but also questions the validity of the respective observational procedures.

The problem may be resolved more effectively by presenting the data available and potentially possible as a system (data bank) with integration bases of its own, differing from the existing disciplinary preferences. These bases, however, cannot be either inductively formulated or pre-set in a purely deductive manner: as a logical problem, the question of many-to-one relationship with its long and eventful history has not been disposed of to this day (for its explication and critical retrospective (see, for instance, [17] and [18]). Therefore, the perspective of obtaining primary empirical data meeting the criteria of relevance, comparability and representativeness is also limited to the existing, historically formed and at least partly realized conceptions and approaches.

Of particular interest is the approach, sketched by the American science historian Kuhn [5] and substantially developed in contemporary studies in the sociology of science. This approach has been successfully and repeatedly used to reveal and study factors determining and governing knowledge generation processes (see, for instance, [21] and [22]), and during the past decade was frequently viewed as a basic methodological pattern to guarantee an integrated multidimensional description of a certain category of phenomena. This paper addresses itself to an analysis of the constructive means of empirical data treatment, implicit in Kuhn's conception of the history of science.

THE CONCEPT OF A PARADIGM AND THE PROBLEM OF RESTRICTING THE OBSERVATION FIELD

In my view, Kuhn's approach was largely motivated by the specific features of empirical data that a historian of science usually deals with. These data are in most cases heterogeneous, contradictory and sometimes even unreliable, and as a result, their treatment as primary data on a certain object is justified only *ex post facto*, in research retrospective. Prior to research, in the perspective of its further development, such treatment is permitted solely "on credit", as a heuristic assumption.

Essentially the same problems arise in an interdisciplinary study of research and it is therefore appropriate to deal with them by using the same instrumentalities as in science history studies. Needless to say, this does not imply that the primary data thus obtained would be based on an arbitrary convention as to assessment criteria or the rules of interpreting empirical data. It is merely suggested that to obtain reliable empirical data it is necessary first to restrict indeterminateness regarding the content of primary data.* In other words, the establishment of invariant characteristics of the totality of phenomena being studied is a pre-requisite for the implementation of relevant interdisciplinary or science history research.

This problem has been explicitly discussed for the first time during the past few decades, mainly in connection with the use of measurement data and mathematical models in politicoeconomic and sociological research. Yet its significance for science is universal for it arises and is somehow tackled in

---

*A good example is the isolation of the corpus of information documents, i.e. texts embodying the results of research. In fundamental research a scientific result is new, heretofore unknown, statements about reality. Therefore, an information document in this case would be a publication or an oral presentation. In applied research and development, however, a scientific result is primarily a mode of action making it possible to produce a certain economic effect: an information document in this case would be a certificate of authorship, a patent or a technical report describing the above mode of action. The boundaries of the corpus thus isolated would therefore vary depending on the initial orientation, and the texts, informative in one respect, may prove to be quite worthless in another [26, p. 117].

constructing any, not necessarily quantitative, phenomenology. A good example is the treatment of material evidence in crime detection science as an integral system: this view represents a legislatively codified convention on the prerequisites of crime investigation rather than an empirically founded substantive statement. It is no coincidence that an eyewitness testimony where the phenomena observed (and consequently statements about them) are presented in syncretic unity has been traditionally viewed as the most highly respected historical source and legal argument.

First of all let us examine in this connection the principles underlying the identification of phenomena observed in studies of the history of science, more traditional than [5]. These studies generally assumed that a historical whole under examination is localized in space and time and that a boundary in one and the other is the criterion making it possible to identify the object of study. This, in part, is the principle realized in [28] where the totality of phenomena under observation is broken down into a number of discrete relatively closed observation areas (the "science of the antiquity", or "medieval European science") limited to the beginning and the end of the respective period or region. The principle underlying the isolation and primary description of phenomena in this case is reference to their order and place (cf. the division of a monograph into chapters), i.e. those phenomena are regarded as close to each other in character which are close to each other in space and time. The principle has been unswervingly observed in all academic work on history of science and it determines to a great extent the division of this discipline into areas of research. Indeed, historians of science traditionally define themselves as specialists in certain periods and regions whereas additional thematic divisions are superposed on this initial taxonomy (for details see [42] giving a fuller treatment of the subject).

One cannot go into detail discussing the exceedingly complex issue involved in the function of temporal and spatial concepts in research on history of science (or, still less, in its other dimensions). Suffice it to say that at least in some important cases the principle of topo- or chronological differentiation or identification of phenomena proves inadequate and, as a result, the observer either has to record all phenomena, presented in a pre-set area, without any exception (which is simply impossible even with the most meticulous attention to the business at hand) or to introduce new constructive devices in addition to space and time. Such is, for instance the function of socioeconomic constructs in [28] where differences in the mode of institutional and financial support are used to clarify the boundaries between certain categories of phenomena.

This situation usually arises when the boundaries of the totality of phenomena under study prove fluid and dependent on conventions concerning premises and observational procedures. It will be recalled that such a situation arose at the turn of the century in some areas of physics and biology. Yet an explication of constructive devices to handle empirical data first promoted a critique of the classical ("substantial") concepts of time and space and then helped form (though by no means completely) the so-called relational conception of the physical continuum, according to which the order of phenomena in time and space is a function of causal relations between them [12]. In social sciences the discussion of these problems began not so long ago [14, pp. 79-120] and therefore the problem of identifying the phenomena being observed is still non-trivial. At any rate the numerous attempts to develop a formalized procedure for the assessment of scientific achievements (i.e. at least, to introduce criteria permitting scientific knowledge to be distinguished from other types of phenomena) pinpoint the problem in a sufficiently explicit form (see, for instance, the discussion of this issue in [4].

Biographical circumstances pushed the problem of identification into the foreground in Kuhn's conception. Kuhn's point of departure was that the "historical integrity" he scrutinized was a limited totality of scientific theories or other cognitive phenomena with some intuitively apparent dissimilarities between them (such are, in Kuhn's view the differences between pre- and post-Copernican cosmologies, as well as physical and sociological theories). Besides, it is assumed that these differences may be formulated and listed by a logically non-contradictory universal rule (this is needless to say, a recapitulation rather than a verbatim outline of the relevant theses). These simple and sufficiently plausible assumptions enable Kuhn to view any phenomena as representing the object of his study provided the differences between them meet some pre-set restrictions. These may be (albeit not necessarily) restrictions in form and/or time; what is suggested here is that the criteria of space and time are insufficient to identify phenomena in a science historian's scrutiny and not that these phenomena are void of duration or extent.

In other words, Kuhn's conception of history of science is based on the principle that there is a relation of homology between phenomena of the same class. This relation is assumed to be invariable regardless of any transformations in the isolated totality of phenomena (for instance, with new ones discovered or those previously explored eliminated) and is the constitutive feature of an ordered, unique and potentially infinite totality of phenomena - theories, individual statements, methodologies, etc. This approach requires in practice that the researcher should spell out or postulate the existence of a certain stable dependence between the variables characterizing a given totality of phenomena, as for instance, between the frequency of the key words and the age of the text (a method used in the attribution of historical sources and literary works). Whenever such a dependence is spelled out (e.g. in the form of a matrix or explicit function), the phenomena being observed are thought to be interchangeable and in this sense identical to a representation of some integral object. It is the relation, projected to a set of phenomena being observed (in Kuhn's terminology a paradigm) that constituted the initial and most general premise for restricting the observation area.

In this function the concept of a paradigm, as proposed by Kuhn, is evidently related to that of a structure in Levi-Strauss' inquiries into the methodology of a myth study [40]. It should be remembered that myths are usually recorded in later versions, in a fragmentary fashion and contaminated by other myths. Therefore, the initial stage of their studies poses essentially the same question as the one under discussion: what are the ways of finding out whether a certain myth record is complete or authentic in the absence of anything but some other records whose completeness and authenticity also give rise to legitimate doubts?

In reply to this question Levi-Strauss first of all demonstrates that any myth is represented by statements concerning some elementary situations or events and that similarity or dissimilarity can always be established between these situations ("motives"), among other things, as regards their place in the text of the myth. This assumption makes it possible to transform the corpus of myth records so that it could be regarded as a potentially infinite totality of statements, constructed in accordance with some immanent rules of its own. Thus the myth under examination is thought to be a specific object while the local myth records are treated as phenomena representing the object in observational situations.

It may be easily seen that in this case, just as in Kuhn's conception of history of science, the identification of statements (as the phenomenology of the object of inquiry) is achieved by restricting the dissimilarities that could exist between the phenomena observed (theories of the same class or records of the same myth). For instance, it is essential for the identification of the myth of Oedipus as an archaic and peripheral version of the myth of Dionysus that the attributes of these people exhibit a relationship of homology which is regarded as a basis for identification. This, in turn, makes it possible to reveal features, specific to this category of people ("archetype" in the sense of [9], in this case somatic defects as well as an extraordinary status in a system of kinship relations).

Of course, such an approach is not the only one possible. Nor does it rule out a more traditional way of restricting the totality of phenomena. Thus the place and order of phenomena in space and time are regularly taken into account in identifying various functional classes of documents (depending on the arrangement of requisites) or legal objects of an arbitrary nature (fingerprinting, ringing, seating arrangements in a conference room, identity card systems) and under special circumstances are regarded as a basis of ideological or ethical evaluation [8]. Nevertheless, identification according to structure has serious heuristic advantages (cf. in this connection the reconstruction in [2] of the proto-Slavic calendar myth) and can probably be used in studying a sufficiently broad class of non-physical phenomena (cf. for instance, in [9] an attempt to use a similar procedure as the basis of a general theory of classification. At any rate, this approach makes it possible to apply observational procedures and indices without resorting to assumptions about the extent or age of a given totality of phenomena [57].

THE CONCEPT OF SCIENTIFIC CHANGE AND THE PROBLEM OF SELECTING OBSERVATION UNITS

Precisely what phenomena possess the appropriate invariant characteristics (these phenomena will be referred to as observation units)? In the established fields of research the nomenclature and principles of registering such phenomena are imposed by tradition and usually are not subject to discussion: a sociologist will not hesitate to define as primary data statements concerning the respondents' reactions without pausing to reflect whether these reactions are direct or whether they are mediated by a certain convention. When the researcher can rely on an established professional tradition, such questions are resolved on the basis of intuition and are rightly considered improper as testifying to lack of professional training.

Entirely different is the situation in nascent research areas where such a traditional technique of description is either lacking or insufficiently developed. Thus, for instance, a mere extension of linguistic procedures to arbitrary sign systems required a preliminary discussion of premises for a study of such systems and, consequently of the conceptual content of meaning and text [7]. Still more complex situations arise in interdisciplinary studies where several independent traditions coexist from the outset [33] and therefore the reality under study acquires several alternative representations. It is no coincidence that research studies inevitably involve controversies about the procedures and methods guaranteeing the most adequate description of the phenomena being observed [4, 49].

Perhaps, the most elaborate attempt to formulate the principles underlying the selection of observation units was made within the programme of logical positivism. It so happened that in the course of its implementation serious discrepancies were revealed between empirical data obtained in different theoretical or procedural contexts. It was correctly assumed, however, that difficulties arising in comparing data of different origin stem from discrepancies in observational procedures and may be eliminated by isolating procedures, invariant to specific conditions of observation. It was, among other things, on these grounds that at least some of the programmatic guidelines of logical positivism were formulated, including, in the first place, the postulate concerning the reducibility of any statements about an assigned totality of phenomena to a few basic ones.

Indeed, statements characterizing a certain objectively homogeneous totality of phenomena must be, first of all, logically compatible. The truth of one of them should not imply the falsehood of another. On the other hand, lack of compatibility may signify that the statements being reduced refer to heterogeneous phenomena, that each of the statements pertains to a different description level, that at least one of the statements being reduced (it is not clear which) is wrong and, finally, that the attempts at reduction were based on improper means. As an initial assumption one might as well consider any of the above alternatives whereas the assessment of non-reduced statements as pertaining to homogeneous phenomena turns out to be just one out of many possible alternatives, and what is more, requiring a special rationale. Therefore, a description of any totality of phenomena (whether it is a theory or just a specification of features) to a first approximation can always be regarded as a system of statements extensionally connected by implications of truth values.

In such a system, however, there will always be at least one statement whose truth (or falsehood) is an independent variable and, consequently, characterizes the direct relationship of the system of utterances to the phenomena being observed. Such statements are considered basic ones, reduction to which (i.e. presentation of their truth as a condition for the truth of any other statements, making up the description) makes it possible to identify statements which are clearly not elements of this description.

Basic statements as such are not given in advance. They acquire the status of variables solely in studying or modelling a description. Therefore, what is essential for defining a statement as basic is, above all, its place in the system, the function it performs, or in other words, pragmatics [19], whereas the mode of producing the statement, its logical structure and other syntactic characteristics are not of decisive importance. This is one of the reasons why relations between truth values may be regarded as models, adequately representing the actual relations between the statements while substantive scientific theories may be consequently likened to constructive objects of mathematics and mathematical logic.

It is in this vital area that the observational methodology, developed within the programme of logical positivism, shows its inadequacy. On the one hand, it puts forward the well-founded thesis to the effect that statements concerning a homogeneous totality of phenomena may be extensionally reduced to some basic statements and that it is the latter that constitute the corpus of primary empirical data. Indeed, "it is hardly possible to develop a theory of meaning for some expressions without assuming that the meaning of other expressions is already comprehensible to us or may be elucidated" [25, p. 9]. On the other hand, the programme of logical positivism virtually proposes an incorrect interpretation of basic statements since it views as such any statements, obtained in a certain privileged manner (so-called assertions about directly observable phenomena), but not necessarily related to some known fixed references. Meanwhile, they must be of necessity statements about some category of phenomena: otherwise any description would have to be regarded as a semantically self-contained axiomatic system which would be an obvious and crude idealization. Therefore, the question as to what should be considered basic statements ("elementary facts" as Russell called them) largely depends on which phenomena are regarded as observational units.

The programme of logical positivism answers this question as follows: observational units are in all cases physiological reactions of the observer regardless of the content of such reactions and, consequently, of the character of phenomena being observed. As is well known, this position proved to be extremely vulnerable. Completely heterogeneous statements, incompatible not only in their logical structure but also in the mode of their production and justification, had to be viewed as empirical data [22]. This, in turn, not only testifies to the limitations of the epistemological premises of logical positivism but also - and this is of fundamental importance - confirms the conventionality or the "conceptual load" of observational standards and procedures. This conclusion is also supported by a comparison of measurement tools and experimental procedures with routine or pre-scientific devices and observational procedures.

A different approach to obtaining and utilizing empirical data was realized by Kuhn in constructing his conception of history of science. His solution of the problem is to spell out precisely what is observed or can be observed in general in a study of research. In fact, the generation of knowledge is, strictly speaking, not a "fact of sensual perception", i.e. observed in the proper sense of the word (as distinguished from, say flashes of light or other optical phenomena, traditionally cited as examples of physical events). Therefore, statements about such phenomena are likewise mediated, knowingly and right from the start, by certain observational techniques [23]. A paradigm was apparently viewed at first as a "device" to enable the researcher to determine and explicate the differences between cognitive phenomena observed (cf. for instance, the use of the term in [43]. As time went on, however, the paradigm acquired a semiotic, sociological and psychological meaning of its own and came to be regarded as a generic term for any restrictions, imposed on differences between homogeneous phenomena. These are restrictions, determined by language semantics, rules of logic, rhetoric and grammar, recognized professional norms, effective legal statutes, established psychological stereotypes and other regulatives of research.

Indeed, it would seem natural to regard each theory, covering a given totality of phenomena, as a transform of a conceptual system, more stable than an individual theory. It is only in this case that a totality of phenomena, isolated by pinpointing their invariant structure, may be thought of as an actual object, a reality rather than an arbitrarily constructed artefact.

This viewpoint is actually a projection of the procedure, used by Kuhn to restrict the observation area, and should therefore be regarded merely as a convenient heuristic assumption. Nevertheless, it makes it possible to indicate the condition to be met so that statements about phenomena whose localization in space and time is uncertain or problematic would represent one and the same integral object. The condition, apparently, is that the phenomena being observed, as, for, instance, records of a myth, should correspond to certain not directly observable changes of the reality studied. The phenomena that can be interpreted as symptoms of such changes are viewed as observational units.

It may be seen that Kuhn's conception designates a specific category of
phenomena as observational units which are treated as being directly
observed, provided we accept the conventions proposed about the premises and
conditions of observation. This approach is a fairly wide-spread solution
to problems, calling for the standardization of initial empirical data
(cf. in this connection the role of the concept of "identical self-
reproduction" in studies of the theory of evolution [20] or the concept of
exchange in contemporary sociopsychological conceptions of behaviour) and
therefore merits a special discussion. At this point, however, suffice it
to say that the standardization of basic statements in content (and not in
their logical form as envisaged by the programme of logical positivism)
enabled Kuhn to avoid the logical circle, common to such situations [24] and
propose a representative and sufficiently rich phenomenology of research. It
should be noted that the emergence of new scientific theories and even the
clarification of the existing ones is traditionally accompanied by the
publication of reports, announcing this event in one way or another. Therefore research too may be identified with the generation of a continuously
growing body of publications: articles, surveys, monographs and finally "the
textbooks from which each new scientific generation learns to practise its
trade" [5, p. 16].

This evidently justifies the treatment of a publication by historians and
sociologists of science as a primary - and the only authentic - fact [11, 29].
In fact, the interpretation of a publication as a references of statements
concerning knowledge generation processes is borne out by the already adopted
forms of progress reports in science: throughout the world research is not
considered completed until a report has been submitted on project
implementation to a competent audience, an article or a monograph has been
published, etc. What is more, despite the multiplicity of historically
realized structural forms of research it has never had any alternative to a
publication (though not necessarily available to all or embodied in a
written text) as a knowledge objectivisation mechanism [15]. Therefore, the
principle "he is a scientist who publishes scientific papers" is not only a
universal and indispensable imperative for the professional activity of
scientists but also a reliable diagnostic feature to identify the phenomena
being observed. It is the body of publications that turns out in this
context to be the "immediate reality" to which statements about knowledge
generation are related.

THE CONCEPT OF NORMAL SCIENCE AND THE PROBLEM OF THE ADEQUACY OF INITIAL ASSUMPTIONS

Further discussion of the premises on which interdisciplinary studies are
based implies an evaluation of the adequacy of initial assumptions. Indeed,
the treatment of paradigm preservation as a basis for the identification of
the phenomena being observed implies that the production of a scientific
result is, at least in principle, reproducible and predictable: otherwise it
could not be regarded as a change of the same object, as in other instances.*

---

*This implies, among other things, that the measure of association of
individual phenomena with the observation area thus isolated is the
probability of their occurrence provided the appropriate structure is
preserved and the totality of phenomena treated in terms of the theory of
probability is comprehensive and qualitatively homogeneous. This view
enables us to introduce the concept of a properly delimited observation area

(*Cont.*)

In other words, in view of the convention adopted concerning the
pre-requisites of research the production of scientific result turns out to
be a function of paradigm preservation, and as a result, the paradigm itself
is viewed by Kuhn not only as a rule for the transformation of statements
about a certain object (i.e. in the narrow and special sense in which the
term has been used by linguists for many decades) but also as a rule for the
transformation of the very object of study. In other words, Kuhn's initial
assumptions suggest the view of a paradigm as a relationship of determination,
immanent to research, a relationship which, once established, imparts
qualitative homogeneity and unity to cognitive phenomena.

This viewpoint is usually formulated in the sense that cognitive phenomena
are "psychologically equivalent" to the application of some normative
standards [16]. Although this convention suggests some risky analogies, it
is not altogether arbitrary: as is known, the level of scientists'
productivity largely depends on their compliance or non-compliance with
certain professional standards of doing research [4, 7]. This is demonstrated,
among other things, by numerous well-documented examples of scientists'
intellectual isolationism rejecting any theories or observation data, at
variance with the accepted standard of a proper scientific result [50].
Besides, participation in research, at least in present-day "big science",
is the exclusive privilege of these who have received special training and
therefore adhere to certain standards in posing and solving scientific
problems. It is these standards that prove to be the required point of
departure to obtain any scientific result and a substratum of psychic
processes, social relations or science's flows of information [51, 53].

Yet the thesis concerning the determination of research by normatively
imposed structural restrictions performs in this case a certain methodological
function, and it is solely for this reason that it is regarded as a
substantive assertion. Therefore, it makes no difference whether the concept
of a paradigm has some empirical content (psychological or sociological) or
is a logical fiction introduced for a preliminary systematization of science
history data available. The essential thing is that in handling primary
data the assumptions made should be complied with. It is precisely in view
of this requirement that the cognitive phenomena being observed are related
to some non-observed changes in the object of study and viewed as being
generated by a certain "legalized" procedure.

In other words, the phenomena subject to registration and recording are
identified by Kuhn at the cost of normative substantive conceptions regarding
the mechanism, governing changes in the object of study. This, in turn,
explains the motivation behind Kuhn's identification of his "historical
integrity" with a communicative system, based on directly cognizable and
deliberately maintained standards of generating messages. The point is that
the area where the procedure of obtaining acceptable scientific results is
known in advance (or, which is methodologically the same, may be established
by contacting an informant) constitutes, so to speak, the "zero level" of

---

where, by definition, some stable probability distribution obtains and to
estimate the totalities of phenomena under study in terms of the deviation
of their statistical parameters from a pre-set ideal type. This approach
proves to be most effective in estimating the representativeness of different
types of samples (for instance, library collections) and was successfully
used in the recent period to reveal structural differences between various
categories of phenomena [1 , 3].

research. Therefore, the explication of a paradigm as "scientific orthodoxy" [45], i.e. a system of verbalized (at least, in principle) prescriptions and interdictions regarding the content of scientific publications as well as their form and the mode of dissemination, makes it possible to pinpoint a category of phenomena whose identification requires no special justification. This category of phenomena is described by Kuhn as "normal science", and corresponding cognitive phenomena are defined as a "strenuous and devoted attempt to force nature into conceptual boxes supplied by professional education" [5, p. 21].

Such attempts are apparently not always crowned with success, and therefore the problem arises as to the adequacy of the convention adopted. Indeed, as distinct from the structure, determining the universe of relations between the cognitive phenomena observed, "scientific orthodoxy" records solely the relations of a stable routine character, deliberately reproduced by research participants. Therefore, in addition to scientific results, envisaged by the paradigm and initiated by appropriate generative procedures ("puzzle-solving"), the development of science involves unforeseen "anomalies", i.e. results, not presentable as a function of the implementation of the accepted professional standards. These results, however, should also be incorporated into the totality of phenomena observed, and certain rational criteria should be spelled out for their identification (i.e. in terms of the convention adopted concerning observation premises, implicational rules should be given for the occurrence of relevant phenomena). Besides, in addition to "anomalies", knowledge generation is accompanied by "episodes in which ... shift of professional commitments occurs" [5, p. 22]. However, if the shift of the communication standard ("scientific revolution") does not mean disappearance of the communicative system isolated, and emergence of a new one, the observance of the above standard is apparently not the sole and universal basis for identification: otherwise each change in the above standard would imply the isolation of a new totality of phenomena, incommensurable with the original one.

Here, in my view, we are confronted with a problem, revealed also in other modern conceptions of research and determining to a considerable extent the trends of methodological reflection in this area. It confronts, for instance, Lakatos [6], when he demonstrates that the restrictions, imposed on transition from one system of extensionally related statements (the proof of Bernoulli's theorem on polyhedrons) to an equivalent one, are extralogical in character and are not "identically true formulas" of mathematical logic. As long as the above formulas remain valid within each individual proof, we form a picture (with objects of observation diachronically ordered) that differs but little from the one sketched by Kuhn: areas within which transition from one statement to another is effected by means of a standard generative procedure are succeeded by gaps in the continuum, "leaps" to a new ordered sequence of statements and, consequently, a new totality of phenomena [31, pp. 181-316].

In other words, representations of research, obtained by using the concept of a paradigm or its explicata, are obviously incomplete. On the one hand, as we have already seen, such representations involve a sufficiently plausible initial assumption concerning the existence of stable regulatives, determining the occurrence of phenomena being observed. For instance, Kuhn introduces and actively uses concepts concerning attitudes, inherent in the researcher's mind, concerning sufficiently specific mechanisms of social control, functioning in global and local systems of scientific communication and, finally, concerning stable (at least, for the duration of a certain period) standards, determining the content of scientific theories. On the other hand, the maintenance of professional standards or effective

functioning of social control mechanisms prove to be a non-universal (and
consequently non-unique) basis of identification. The results of actual
studies, on which Kuhn bases his argument, testify to the possibility of
spontaneous productive acts (insight) and non-sanctioned information flows
(so-called "informal contacts"), to changes and even inversions [44] in
the accepted communication standards and even to a regular emergence of
unorthodox social groupings and unforeseen results.

As is known, this a point where both "internalists" and "externalists" are
agreed [10] which shows once again the illusory nature of the controversies
between these two approaches to knowledge generation studies. The former
believe that scientific changes arise from scientists' personal ambition,
their disposition to intellectual innovations, dedication to certain research
programmes or other "internal" motives. The latter maintain that the major
- or, at least, the most powerful - stimuli behind scientific changes are the
policy *vis-à-vis* science, human migrations from one field of activity to
another, economic crises or social revolutions, requirements of the current
military-political situation and other "external" factors. In both cases,
however, scientific change is viewed as an arbitrary and essentially
uncontrollable break of order except that the "internalists" locate the
source of disturbance in the scientist's mind and the "externalists" in
social institutions interacting with science. Needless to say, this should
not be understood in the sense that the institution of science or individual
scientific disciplines do not undergo any sudden irreversible changes under
the impact of other social institutions, restricted social groups or even
certain individuals. It is merely suggested that reference to such
influences is unacceptable as a point of departure in analysis.

In other words, compliance or non-compliance with standards, effective
within a scientific communication system, does not always or necessarily
imply actual association or non-association with this system. Indeed, quite
a few situations are known where compliance with the standards adopted in
the course of professional education led to the emergence of various "pseudo-
effects", including conceptions and programmes, built on sand [41], and on
the other hand, explicit and deliberate deviations from professional
standards being applied have to be reckoned with as key facts determining
the interpretation of empirical data [13].

These considerations show - or at least support the conclusion - that in a
study of knowledge generation processes it is necessary in addition to the
"scientific orthodoxy" to take into account other, supplementary bases of
identification. Indeed, the construction of an infinitely varied totality
of statements by a universal generative procedure is impossible in view of
the restrictions, imposed on the completeness of extensionally related
systems by Gödel's theorem [48]. Therefore, the identification of any - and
not only some - phenomena, observed in a study of research implies either
the existence of spontaneous innovations (which is impossible in view of
the assumption adopted as a point of departure for observation) or the
introduction of several independent identification bases making it possible
to isolate certain categories of phenomena, limited in scope. In this
sense Kuhn's well-known proposition concerning the multiplicity and
incommensurability of paradigms is an expression of an antinomy, noted in
one form or another by every analyst of knowledge generation processes.

THE CONCEPT OF A SCIENTIFIC COMMUNITY AND THE PROBLEM OF COMPLETE EMPIRICAL DATA

Let us turn in this connection to the concept of a scientific community whose inclusion in the methodological arsenal of a science history study has been frequently regarded as Kuhn's major contribution to science. This concept is introduced as the explicatum of a paradigm and its application at first did not presuppose the use of sociological substantive concepts or methodological procedures. Kuhn merely takes it for granted that differences between paradigms are directly manifested not only as differences between totalities of cognitive phenomena but also as differences between social groupings of scientists (an assumption later substantiated empirically). This helped to eliminate the deficit of data that characterized some episodes in the history of science, poorly documented by publications [30] and to present the totality of phenomena under study as a continuous temporal series.

At a first glance, this approach looks like a mere extension of the existing conventions making it possible to view as observational units not only publications but also non-written communicative events. Upon a more careful examination, however, it becomes clear that the phenomena of different categories are not equivalent, that they unquestionably form a hierarchy with functional dependence and causal relations between them. Indeed, in Kuhn's view, the meaning of cognitive phenomena is relative and largely depends on the depth of the concomitant or contemplated restructuring of social groupings: "For the far smaller professional group affected by them, Maxwell's equations were as revolutionary as Einstein's, and they were resisted accordingly" [5, p. 23]. At the same time, in Kuhn's conception references to phenomena of institutional control, the formation of intellectual stereotypes or conformity with established professional standards are made primarily to explain the relevant cognitive phenomena, such as the stability of theories, intrinsically contradictory or insufficiently supported by empirical data. In other words, the concept of a scientific community, in fact, not only refers to a certain category of phenomena but is also used as a constructive device.

This shift of the accent from semiotic to sociopsychological determinants of research reflects not only the hierarchy of Kuhn's tools to handle empirical data but also a fundamental evolution of his own views. This becomes abundantly clear after comparing the initial version of his conception, published in 1962, and the special supplement included in the 1970 edition. The original version introduces the concept of a scientific community only sporadically, mainly to fill gaps in the empirical picture presented; the concept proves to be redundant when the idea of a paradigm can be properly explicated in terms of methodology and logic. In the supplement, however, the concept of a scientific community is predominant while logical and methodological explications are used merely for explanations when a purely sociological interpretation of phenomena turns out to be unconvincing or trivial.

The shift in this direction was, in all probability largely due to the papers by Merton's followers, published in 1965-1969 and dealing with the institutionalization of research [56]. These papers, partly stimulated by Kuhn's publications, showed and analysed in detail the significance of consensus among research participants as a major condition for the stability and effective functioning of the appropriate social groupings. This laid the empirical groundwork for Kuhn's initial assumption concerning a functional relationship between the application of certain communication standards and knowledge generation processes.

As a result of such clarification, Kuhn's conception assimilated a series of disputable and even dubious substantive propositions which easily lend themselves to extreme conclusions. However, what we are concerned with is first and foremost whether reference to a scientific community helps to guarantee the completeness of empirical data and, only to that extent, whether the theoretical constructs to be used in such reference are acceptable. This, in my view, is the actual borderline between a sociological study of professional groupings and an interdisciplinary study of activity where sociological concepts are drawn upon as a constructive device. I shall therefore try to find out what methodological function is in this case assigned to the concept of a scientific community without discussing any alternatives to Kuhn's approach.

In replying to this question it is necessary, in the first place, to ascertain the precise empirical content attributed to this concept. It should be noted that the papers analysing contemporary conceptions of research usually do not distinguish between a scientific community and other social groupings of scientists. Meanwhile, academic American sociology traditionally uses the term "community" in the sense of such, and only such, a social grouping whose identification is based on informal personal relations between its members. In other words, the concept points rather to a specific mode of determination of communicative processes (in this sense, the concept of a community is equivalent to that of a social network). Communities are invariably contrasted to formal social groupings (based on depersonalized normatively assigned behavioural standards). This dichotomy is contained in any serious guidebook or textbook on sociology and the concept of the existing terminological norm. In other words, regardless of Kuhn's motivation in introducing this concept, the description of research associations as communities actually amounts to the adoption of sufficiently specific assumptions as to knowledge generation processes [54].

This excursion into etymology may be supported by other arguments. If a scientific community is viewed merely as a totality of paradigm representatives, it is obvious that it is the concept of a paradigm that bears the methodological brunt in a study of research (as it happened in [38] or [55]) and one can get along without using any sociological theoretical constructs whatsoever. On the other hand, if the concept of a paradigm is covered by that of a scientific community, then the paradigm becomes an epiphenomenon of professional communication whereas it is social relations that become the actual determinants of scientific change [47, 50]. This is one of the reasons why the treatment of a paradigm and a scientific community as related concepts implies in perspective either the elimination of psychological, social, economic and other "external" conditions of research or, on the contrary, the reduction of cognitive phenomena to social ones. In this sense the 1969 supplement where Kuhn partly agrees with the above interpretation is a step back in relation to the original version of the conception.

Thus the concept of a scientific community is not so much an explicatum of the concept of a paradigm as its complement in the sense that scientific change is independently determined by both language paradigms and interpersonal relations. This conclusion, widely discussed in contemporary West European sociology of science, explains the actual significance acquired by sociological constructs and procedures in Kuhn's conception of the history of science. In his view, both the concept of a paradigm and the concept of a scientific community refer to communication standards. Their realization may be regarded in an equal degree as a basis for the identification of phenomena being observed.

At the same time, the above standards differ from each other – at least, in the mode of their realization – and the phenomena isolated by using the corresponding methodological procedures are not equivalent to each other. While the concept of a paradigm, imposing initial assumptions on the object of study, permits the isolation of phenomena regarded as being directly observed, the concept of a scientific community, imposing additional assumptions on the same object, permits the isolation of a category of phenomena whose identification calls for a special analysis. Indeed, for Kuhn reference to social relations is in fact a means of identifying scientific results whose production involves deviations from the existing paradigm: it is only in such special cases that Kuhn introduces and actively uses the notion of competition between different research associations or of group solidarity between representatives of the same professional category. The most fundamental methodological load is assigned to the community concept in describing scientific revolutions. The dynamics of the latter are presented by Kuhn solely in terms of interpersonal and intergroup contacts.

Therefore it is clear that the concept of a scientific community is used by Kuhn as a constructive device, additional to a paradigm and applied to resolve antinomies which arise in identifying the phenomena observed in terms of structure and function. At a first glance this approach seems to be most effective and it did arouse great enthusiasm among scholars exploring the determination of cognitive processes (cf. for instance, the evaluation of the heuristic potential of sociological concepts in the editors' commentary to [32]). Nevertheless, the results obtained in a study of interpersonal relations in research associations proved to be less promising and at least not measuring up to the hopes, originally pinned on the relevant concepts.

The fact is that the function, performed by interpersonal relations in regulating research, turned out to be essentially different from what was originally expected. As we have seen, in Kuhn's view, interpersonal relations are the mechanism ensuring and accounting for a change in the existing professional standards, and in this sense they complement "scientific orthodoxy". Yet numerous sociological studies indicate that functionally interpersonal relations are coincident with "scientific orthodoxy" as they ensure and account for its stability. What is more, it is precisely interpersonal relations that guide and regulate research in situations where appropriate standards are lacking, challenged or insufficiently developed [46].

In the final analysis, the concept of a scientific community duplicates the concept of a paradigm, and its introduction, at best, gives an additional dimension to "puzzle-solving". As to the identification of "anomalies" and "scientific revolutions", the function of interpersonal relations in this case appears very problematic and unclear. It may be probably maintained that the established network of interpersonal relations determines to a considerable extent migration flows between research associations and also regulates the adaptation and readaptation of scientists. However, the question as to whether interpersonal relations are a factor, conducive to a change in professional standards rather than individual features of communication systems, remains unresolved.* Intensive empirical and theoretical studies are under way in this direction. Their premises and results require a separate treatment.

---

*Mention should be made here of the approach to a study of changes in communication systems, proposed in the twenties by philologists, influenced by Marr. There was a time when this approach was instrumental in forming some methodologically remarkable linguistic and literary conceptions (albeit

In other words, while Kuhn's convention concerning premises for the acquisition and use of empirical data permit knowledge generation processes to be represented in a holistic and multifaceted manner, it also leads to some fairly specific methodological problems. For his part, Kuhn merely points out periodic changes in professional standards of research without giving them any rational interpretation and, consequently, without any special reference to constructive devices to guarantee a reliable identification of such phenomena (for instance, discrimination between "anomalies" and clearly unacceptable scientific results). Meanwhile, the incompleteness of Kuhn's empirical picture calls in question not only the adequacy of his conceptions, but also - which is more important - determination itself and consequently the possibility of forecasting structural changes or controlling them (see for instance, an evaluation of Kuhn's conception in [26, pp. 80-81]). It is no coincidence that Kuhn himself uses in the title of his monograph the terms "structure" (a principle that should be held in identifying the phenomena observed) and "scientific revolution" (a category of phenomena requiring special devices for its identification).

CLOSING REMARKS

We have seen that an analysis of Kuhn's conception shows that it is based on fairly specific methodological principles that permit the handling of problematic and heterogeneous empirical data. These principles and the constructive devices developed on their basis are coextensive, to a first approximation, with the set of concepts and procedures, long and widely known as "structural analysis", somewhat modified for an interdisciplinary study of research. Such modification, however, permitted the dynamics of scientific knowledge to be represented in a non-trivial manner thus over- coming the difficulties involved in an objective and uniform description of this class of phenomena.

In perspective, the principles of structural analysis may prove to be an effective tool for the integration of observational indices and procedures, used to obtain information on knowledge generation processes. Until recently such integration was effected, in the main, by limiting the totality of phenomena under study to a single social grouping or body of publications with a subsequent extrapolation of patterns revealed to other observation areas. This strategy, however, tends to absolutize features, inherent in a strictly determinate category of phenomena, and, consequently, the representativeness of the empirical data obtained remains in such cases a moot question. The approach, discussed in this paper, makes it possible to abstract oneself from local structural features of research and introduce a generalized representation of research as a stable totality of functional relations. This activity is described through a universal set of indices. Such representation, while guaranteeing unlimited comparability of information on knowledge generation processes, would be a useful instrument of decision-making in research management.

---

not unquestionable in content) and is now being actively revived in semiotics. From this standpoint, the dynamics of communication systems comes from the interaction of at least two hierarchically ordered structures, one of which, represented by dominant universally binding standards, ensures the reproduction of the system, while the other represented by ousted marginal preferences, ensures the assimilation of changes. As is known, a hierarchy of structures, similar in their functions and mode of realization, is set up in sociological studies of mass media and "the leading edge" of research [36, 39].

Interdisciplinary Study of Scientific Activity 207

Needless to say, the implementation of the principles of structural analysis by Kuhn himself as well as by some of his followers among American and West European historians and sociologists of science gives rise to serious objections. We have been mainly concerned with a clarification of the assumptions, made by Kuhn as premises of his own research and therefore did not discuss at all the way he adheres to the initial convention. There is a vast amount of literature, devoted to a critique of Kuhn's conception. I shall merely name its major trends: interpretation of the concepts of a paradigm and normal science (their vulnerability was already noted in [38]), motivation behind the use of data from history of science (it appears at times as a way of getting round the difficulties, involved in a structural interpretation of knowledge generation processes), the status of sociological concepts in a study of cognitive phenomena (partially touched upon in this paper), and finally, epistemological problems arising in view of the initial assumptions. But as far as I am concerned the main thing is that Kuhn's initial methodological orientation is not consistently realized, and as a result the problems, arising in an interdisciplinary study of research do not receive an elaborate and adequate treatment. This in my view gave rise to programmes, actively supported by Kuhn, for independent studies of linguistic and social determination of research which in effect was tantamount for compartmentalization of his original programme into isolated disciplinary units.

The approach I am discussing can develop – and is in fact, developing to some extent – in several contiguous directions. First, there is a need for further efforts to explicate the principles of structural analysis as specialized observational procedures. This explication can probably be based on a totality of methods, elaborated to reveal quotation relationships in a body of scientific publications as well as social networks of science. The number of studies where such methods are being used to explore the dynamics of scientific knowledge is now rapidly growing thereby showing how important these problems are.

Second, the problem of relating the assumptions and constructs of structural analysis to the present-day organizational structure of research is still the order of the day. It so happens that the empirical data, characterizing the dynamics of scientific knowledge, are obtained primarily by a study of provisional social groupings ("scientific schools", "invisible colleges" and "coherent groups"), arising and maintained solely by virtue of the personal interest of their members in the maintenance of contacts. Meanwhile a sizable part of the total amount of research is carried out by permanent research institutions whose functioning implies special efforts for the coordination of research and the existence of a full-fledged administrative staff. Therefore, further steps toward an interdisciplinary study of research presuppose an explication of structural restrictions, regulating knowledge generation processes, in terms of technical, legal and economic norms (budget, research project, authorized staff, equipment, personnel, relations of authority). Some preliminary steps have already been taken in this direction (see an attempt in [21] to use structural analysis constructs for this codification of the established practice of financing fundamental research). Yet the problem has not yet been fully formulated in all its magnitude.

Third, we have seen that the use of the constructs and procedures of structural analysis to study knowledge generation processes, leading as it does, to a successful solution of the existing problems gives rise to new ones. In particular, the emergence, of at least some categories of cognitive phenomena involves deviations from the existing professional standards of research thus making these phenomena inaccessible to classification in accordance with the established criteria. The methods, currently used for the identification of structural changes, are based primarily on explicit or implicit recourse to expert judgements, for example, in regard to relations of continuity between research areas and disciplines. Yet such judgements reflect, at best, a subjective awareness of kinship or a gap between juxtaposed research standards, and therefore need a systematic adjustment to the character and degree of actually observed differences. This, in turn, suggests the need for a special study of mechanisms behind structural changes as well as constructive devices for an isolation and study of corresponding relations and processes.

REFERENCES

1. M. V. Arapov and A. N. Libkind, On the concept of a closed information flow, *NTI* Ser. 2, No. 6, 1-15 (1977), in Russian.
2. V. V. Ivanov and V. N. Toporov, On the problem of the reliability of late secondary sources in connection with studies in mythology, in: *Works on Semiotics*, Vol. 1, pp. 46-82, Tartu, Transactions of Tartu University (1973), in Russian.
3. A. A. Ignatiev and A. I. Yablonsky, Analytical structures of scientific communication, in: *Systems Research Yearbook 1975*, pp. 64-81. Moscow, Nauka (1975), in Russian.
4. V. Zh. Kelle, Methodological problems in a multidimensional study of research, in: *Problems in the Activity of a Scientist and a Scientific Team*, pp. 26-39. Moscow, Nauka (1979), in Russian.
5. T. S. Kuhn, *The Structure of Scientific Revolutions*. Chicago, University of Chicago Press (1970).
6. I. Lakatos, *Proofs and Refutations*. Moscow, Nauka (1967), in Russian.
7. Yu. M. Lotman, On the problem of meaning in secondary modelling systems, in: *Works on semiotics*, Vol. II, pp. 22-37. Tartu, Transactions of Tartu University (1965), in Russian.
8. Yu. M. Lotman, The conception of geographical space in medieval Russian texts, *ibid.*, pp. 210-216, in Russian.
9. S. V. Meyen and Yu. A. Shreyder, Methodological aspects of classification theory, *Voprosy Filosofii* No. 12, 67-79 (1976), in Russian.
10. S. R. Mikulinsky and L. A. Markova, On different interpretations of motive forces in the development of science, *Voprosy Filosofii* No. 8, 107-117 (1971), in Russian.
11. E. M. Mirsky, Publication array and the system of a scientific discipline, in: *Systems Research Yearbook 1977*, pp. 107-117. Moscow, Nauka (1977), in Russian).
12. Yu. B. Molchanov, *Four conceptions of time in philosophy and physics*. Moscow, Nauka (1977), in Russian.
13. E. Panofsky, Galileo as a critic of the art (aesthetic attitude and scientific thought), *Isis*, 47, 3-15 (1956).
14. W. Platt, *Strategic Intelligence Production, Basic Principles*. New York (1957).
15. D. J. de Solla Price, Communication in science: the ends - philosophy and forecast, in: *Communication in Science*, pp. 199-209 (A. de Reuck and J. Knight Eds), Boston, Little, Brown and Co. (1967).
16. M. S. Rogovin, The subject-matter and theoretical foundations of cognitive psychology, in: *Foreign Studies in Cognitive Psychology*, pp. 62-149. Moscow, INION AN SSSR (1977), in Russian.
17. G. A. Smirnov, Toward a definition of an ideal integral object, in: *Systems Research Yearbook 1977*, pp. 61-85. Moscow, Nauka (1977), in Russian.

18. G. A. Smirnov, On the initial concepts of a formal theory of integrity, in: *Systems Research Yearbook 1978*. pp. 53-69. Moscow, Nauka (1978), in Russian.
19. V. S. Stepin, On the problem of the structure and genesis of a scientific theory, in: *Philosophy. Methodology. Science*, pp. 112-113. Moscow, Nauka (1972), in Russian.
20. N. V. Timofeyev-Resovsky, The structural levels of biological systems, in: *Systems Research Yearbook 1970*, pp. 80-91. Moscow, Nauka (1970), in Russian.
21. J. G. Wirt, A. J. Lieberman and R. E. Levien, *R & D Management*, Lexington, Mass., Lexington Books (1975).
22. V. S. Shvyrev, Toward an analysis of theoretical and empirical categories in scientific cognition, *Voprosy Filosofii* No. 2, 3-14 (1975), in Russian.
23. V. S. Shvyrev, On the specific character of social cognition, *Voprosy Filosofii* No. 5, 118-128 (1977), in Russian.
24. E. G. Yudin Toward the analysis of the internal structure of generalized systems conceptions, in: *Problems of Systems Research Methodology*, pp. 433-453. Moscow, Mysl' (1970), in Russian.
25. S. A. Yanovskaya, Preface, in: *R. Carnap, Meaning and Necessity*, pp. 5-21. Moscow, Izd-vo Inostr. Lit. (1959), in Russian.
26. E. Jantsch, *Technological Forecasting in Perspective*. Moscow, Progress (1970), in Russian.
27. B. Barnes, The comparison of relief-systems: anomaly versus falsehood, in: *Modes of Thought* (R. Horton and R. Finnegan Eds), pp. 182-198. London (1973).
28. J. Ben-David, *The Scientist's Role in Society: A Comparative Study*. Englewood Cliffs (1971).
29. D. E. Chubin, The journal as a primary data source in the sociology of science: with some observations from sociology, *Social Science Information*, 14, 157-168 (1975).
30. H. M. Collins, The TEA sets: tacit knowledge and scientific networks, *Science Studies* 4, 165-186 (1974).
31. I. Lakatos and A. Musgrave (Eds), *Criticism and the growth of knowledge*. Cambridge (1970).
32. K. D. Knorr et al. (Eds). *Determinants and Control in Scientific Development*. Dordrecht, Boston (1975).
33. G. Frey, Methodological problems of interdisciplinary discussions, *Ratio* 15, 161-182 (1973).
34. G. N. Gilbert, Competition, differentiation and careers in science, *Social Science Information* 16, 103-123 (1977).
35. G. N. Gilbert, Measuring the growth of science, *Scientometrics* 1, 9-34 (1978).
36. W. O. Hagstrom, Competition in science, *American Sociology Review* 39, 1-18 (1974).
37. L. L. Hargens, Relations between work habits, research technologies and eminence in science, *Sociology of Work and Occupation* 5, 97-112 (1978).
38. R. Klima, Scientific knowledge and social control in science: the application of a cognitive theory of behavior to the study of scientific behavior, in: *Social Processes of Scientific Development* (R. Whitley ed.), pp. 96-112. London (1974).
39. J. Law, The development of specialties in science: the case of X-ray protein crystallography, *Science Studies* 3, 275-303 (1973).
40. C. Levi-Strauss, *Structural Anthropology*. New York (1967).
41. G. Magyar, "Pseudo-effects" in experimental physics: some notes for case-studies, *Social Studies Science* 7, 241-267 (1977).
42. C. A. McClelland, Systems and history in international relations - some perspectives for empirical research and theory, in: *General Systems*, Vol. 3, pp. 221-248. Ann Arbor (1958).

43. R. K. Merton, Paradigm for the sociology of knowledge, in: *R. K. Merton. Sociology of Science: Theoretical and Empirical Investigations*, pp. 7-40. Chicago (1973).
44. I. I. Mitroff, Norms and counter-norms in a select group of the Apollo-Moon scientists, American Sociology Reviews, 39, 579-595 (1974).
45. M. J. Mulkay, Some aspects of natural growth in the natural sciences, *Social Research* 36, 22-52 (1969).
46. M. J. Mulkay, Sociology of scientific research community, in: *Science, Technology and Society: A cross-disciplinary Perspective*, (I. Spiegel-Rösing and D. Price de Solla Eds), pp. 93-148. London, Sage (1977).
47. N. C. Millins, The development of a scientific specialty: the phage group and the origin of molecular biology, *Minerva* 10, 51-82 (1972).
48. E. Nagel and J. Newmann, *Gödel's Proof*. New York (1968).
49. G. Radnitzky, From logic of science to theory of research, *Communication and Cognition* 7, 61-124 (1974).
50. Rejected knowledge. *Sociology Review Monographs* Vol. 23 (R. Wallis Ed.) Keele (1978).
51. B. F. Reskin, Scientific productivity and the reward structure of science, *American Sociology Reviews*, 42, 491-504 (1977).
52. D. Robbins and R. Johnston, The role of cognitive and occupational differentiation in scientific controversies, *Social Studies Science* 6, 349-368 (1976).
53. W. van Rossum, The development of sociology in the Netherlands: a network analysis of the editorial board of the Sociologische Gids, in: *Social Processes of Scientific Development*, pp. 172-192. (R. Whitley Ed.), London (1974).
54. W. van Rossum, The community structure of science, in: *Paper, Presented on the Conference of the Research Committee on the Sociology of Sciences*. Budapest, ISA, (1977).
55. W. Stegmüller, Structures and dynamics of theories: some reflections on J. D. Sneed and T. S. Kuhn, *Erkenntnis* 9, 75-100 (1975).
56. N. Storer, Introduction, in: *R. K. Merton. Sociology of Science: Theoretical and Empirical Investigations*, pp. xi-xxxi. Chicago (1973).
57. R. Whitley, Umbrella and polytheistic scientific disciplines and their elites, *Social Studies Science* 6, 471-497 (1976).

THE DEVELOPMENT OF SCIENCE AS AN OPEN SYSTEM

A. I. Yablonsky

Knowledge of the dynamic characteristics of science is essential not only to theoretical research but to the implementation of scientific policy, for science cannot be planned or forecast without some conception of its development. Therefore students of science increasingly resort to the historicogenetic approach which strives to establish the "logic" of scientific development rather than draw up the chronology of scientific discoveries. It is only natural therefore that special emphasis is laid on qualitative and formalized modelling of the development of science.

This paper presents an approach to modelling the development of science on the basis of the current thermodynamic theory of open systems wide of the equilibrium state [5]. The concepts of this theory and a number of additional analogies are then used to construct some mathematical models of the operation of a scientific community.

IRREVERSIBLE PROCESSES AND THE EVOLUTION OF SCIENCE

The Evolution of Open Systems

Irreversible processes occurring in non-equilibrium systems have until recently been considered in a linear approximation, i.e. for states that are close enough to equilibrium to hold linear relations between fluxes and flows [16]. For this case Prigogine has proved the principle of minimum entropy production [6] which ensures the stability of steady states close to equilibrium. In his latest studies [5, 17] he has obtained a number of entirely new results which generalize the "linear" theory so as to embrace the nonlinear case, i.e. systems that are far from the equilibrium state and do not allow of linear approximation. The essence of these results, which characterize the development of various complex systems is discussed below.

Prigogine regarded a system's development as a transition to a steady state with a lower entropy (i.e. a rise in the system's organization). For a system in the equilibrium state or a steady state close to equilibrium (admitting of linear approximations) development was shown to be blocked by "over-stability". Hence, according to Prigogine, the development of a system is only possible in steady states wide of equilibrium which give rise to nonlinear effects; the instability of such states is regarded as a potential source of development.*

Prigogine has demonstrated that far from the state of equilibrium fluctuations do not subside as in stable systems. Instead, they are intensified by some non-linear (e.g. autocatalytic) processes, reach the macroscopic level and cause the system's abrupt transition to a new steady state with a lower entropy. This ordering of the system through fluctuations in unstable states was termed "order through fluctuations". Prigogine has introduced the important notion of "dissipative structures", i.e. structures emerging as a result of such processes.

Order may be achieved in non-equilibrium situations only if there are external fluxes (matter/energy or information) that keep the system far from equilibrium. If there are no such fluxes, i.e. if the system is isolated, these situations generate dissipation processes which destroy the structure and diffuse energy or information, making the system revert to the equilibrium state. By contrast, "open" systems which are closely associated with the environment (e.g. living organisms), as opposed to equilibrium structures like crystals, are capable of development. Interaction with the environment (exchange of matter/energy, influx of information) leads to the emergence of instability followed by a more orderly structure (a new state with a lower entropy).

Dissipative structures are inherent in a broad range of systems, from physico-chemical to biological or even social ones; the latter may also be regarded as open systems whose structure is liable to break down with the interruption of external fluxes (material resources or information) or to evolve if these fluxes are present. The orderliness of ecological systems is a typical instance of a dissipative structure: "the emergence of ecological niches and the spatial distribution of species may be interpreted as the formation of dissipative structures" [21, p. 221]. In the domain of science dissipation may take the form of obsolescence of information, e.g. the exhaustion of the scientific potential of a certain study (manifested in a dwindling number of references to it), causing it to leave the sphere of active scientific interest and join the historical archives of science (for details concerning the obsolescence of information see [40]. It is interesting to note that the obsolescence of information is normally described by the same negative exponential curve e^{-t} as the dissipative processes in other fields: the theory of reliability, radioactive decay, physical chemistry, etc.†

---

*The mathematical relations underlying Prigogine's theory are not shown, for what is needed here is a qualitative outline of the "nonlinear" approach to the development of complex systems. The mathematical side of the matter is presented in [7] and [17].

†The negative exponential curve may be derived from a general theoretical assessment of the reliability (the probability of decomposition) of a system comprising a large number of elements with a fixed probability of failure. This automatically implies that, apart from decomposition (dissipation), each complex system is subject to a counter-process of development, so as, at least, to make up for the dissipation and protect the system from degradation (reversion to the equilibrium state). In the case of science there is experimental evidence of exponential growth.

The Evolution of Open Systems and the Development of Science

The concept of development as a succession of steady states through periods of instability towards higher organization is increasingly used for analysing the dynamics of various systems: physicochemical, biological and others. The most interesting and useful results were obtained by Eigen, who used the thermodynamic theory of open systems to investigate the emergence of "biological" information (self-organization of biological systems) as a result of evolutionary competition and selection [29, 30].

Eigen proceeds from the fact that initially in a biological system information possesses quantity but not value. It acquires value (utility) only in the course of competition and selection among the potential carriers of information, depending upon the reproductive ability of the carrier. What happens is the emergence of a new kind of information, not its mere transfer, as in the communication processes of Shannon's information theory. The process is structurally similar to the conversion of "input" scientific information in the form of separate results etc. into "evaluated" information, i.e. applicable scientific knowledge.

The nonlinear concept of development advanced by Prigogine and Eigen allows certain parallels to be drawn between the evolution of biological and intellectual (Piaget's term [27]) structures. A common approach to the development of science is based on its analogy to biological evolution [13, 47] which may be of interest if one treats this analogy critically. However, we are interested in a more specific comparison of the major postulates of the thermodynamic theory of open systems with some of the current conceptions of scientific development.

One of these is the concept of cyclic development of science through alternation of intensive ("the formation of new conceptual elements" [20, p. 225]) and extensive ("the development of theory within a given conceptual framework" [20, p. 225]) stages. The basic principles of this scheme are commonly known. One of its variants is elaborated by Kuhn [11], who claims that, in a first approximation, science develops spasmodically through scientific revolutions spaced by calmer and longer periods of normal science. During a period of normal science the researchers obtain results on the basis of a common concept or paradigm. The accumulation of accidental discoveries or anomalies that do not fit in the paradigm leads to a crisis of the system. At this stage the normal constraints are weakened, leading to a multiplicity of competitive hypotheses. The scientific revolution produces a new paradigm which marks the end of the crisis and the resumption of normal science.*

---

*A number of Kuhn's assertions were justly criticized by S. P. Mikulinsky and L. A. Markova in their concluding remarks to Kuhn's book [11]. These authors specifically note Kuhn's excessive "algorithmization" of the researcher's activity in the period of normal science (which "should be more correctly termed the period of smooth evolution" [11, p. 277]), and his treatment of the emergence of new knowledge as a mere "choice between the two available theories, old and new, by the scientific community" [11, p. 279]). Besides, Mikulinsky and Markova criticize Kuhn for his violation of the historical approach as manifested in the utter lack of continuity between pre-revolutionary and post-revolutionary science in his theory [11, p. 278].

If one regards this scheme as a possible model of scientific development (of course a highly simplified one), one can trace some similarity between Kuhn's conception of the evolution of science and Prigogine's model for the development of open systems wide of equilibrium. Indeed the periods of normal science may be interpreted as steady states and the paradigms as characteristics of these states (like entropy characteristics in Prigogine's systems). The change of paradigm at the point of crisis (fluctuational instability according to Prigogine) corresponds to the system's transition from one steady state to another, more remote from equilibrium (more highly organized). Finally, the expanded "explanatory" ability of the new paradigm corresponds to a lower entropy and a higher organization of the evolving open system.

One may say that normal science, the scientific process occurring within the framework of one paradigm, is a stable state of science until the "explanatory" ability of this paradigm is exhausted. Then the accumulated anomalies give rise to mistrust of the old paradigm, the existing restrictions become loose, the entropy or diversity of the system increases and the system passes into an unstable state. This is the intermediate divergent stage of the crisis situation, marked by the emergence of a new paradigm. It is followed by an abrupt transition from the old unstable steady state to a new stable one with a lower entropy (a more up-to-date paradigm). This is the convergent stage of the system's transition to a stable steady state at a higher, more progressive level.

The interpretation of instability as a prerequisite for development, i.e. transition to a new state, presupposes two interconnected processes: the negentropic process which keeps the system from degradation and the entropic process which generates the necessary diversity for the emergence of novel characteristics. This assumption accords with the dual nature of scientific activity which may be regarded as (1) the solution of problems (reducing their number and, in a sense, the system's entropy) and (2) generation of problems (leading, in case of their insolubility within the existing paradigm, to anomalies and new hypotheses, i.e. to increasing entropy). In this context one may claim that science is one of the most complicated kinds of human activity that has something in common with random search [32], for its results are to some, often dramatic, extent unpredictable. A scholar is someone who not only reduces the entropy (organizes knowledge) but generates entropy (raises new problems, proposes new lines of research, etc.); however the growing entropy flux is not in this case the conventional synonym of degradation (which would hold true for systems close to equilibrium or for isolated systems) but a source of diversity which, along with the negentropic processes, ensures evolutionary transition to a new state with a lower entropy.

The scientific processes themselves occur threshold-wise, which is also the case, according to Prigogine, with open systems wide of equilibrium. For instance, when modelling the formation of an area of science, Goffman [37] regards the appearance of publications as an epidemic process which successively affects the members of a scientific community, i.e. he applies the approach to the diffusion of novel ideas extensively used in mathematical sociology [39].

Here is the resulting system of differential equations

$$\frac{dS}{dt} = -\beta SI - \delta S + \mu$$

$$\frac{dI}{dt} = \beta SI - \gamma I + \nu \qquad (1)$$

$$\frac{dR}{dt} = \delta S + \gamma I$$

where S, I, R are, respectively, the potential researchers (those who may get their work published), active authors (whose work gets published) and the drop-outs (who have no publications now but had some in the past); β, δ, γ are, respectively, the rates of the emergence of new authors, the "elimination" of potential researchers and the "elimination" of active authors; μ, ν are, respectively, the rates of the influx into the scientific community (as an open system) of new potential researchers or active authors.

It follows from (1) that the system enters the state of an epidemic process if

$$S > \frac{\gamma - \nu/I}{\beta} \equiv \rho \qquad (2)$$

where ρ is the threshold density of potential researchers in the area; an excess over this value is a prerequisite of the epidemic process. Note that the availability of a fresh supply of potential researchers to the area (its "openness" to the external flux) is a pre-condition of the system's development as a series of "recurrent epidemics" [2]: after an outbreak of the epidemic the density of potential authors may drop below the critical level and a relative lull ensues until the threshold is again exceeded due to the incoming researchers, and so on.

To conclude this part of the paper, we would like to point out some interesting parallels between Prigogine's thermodynamic theory of open systems and Piaget's genetic epistemology [15, 27] emphasized by Prigogine himself in his methodological rather than mathematical study [45]. Prigogine compared the irreversible processes in the evolution of dissipative structures to the dynamics of intellectual structures whose organization, in Piaget's conception, increases in the course of gradual equilibration and self-regulation ("progressive equilibrium" according to Piaget [44, p. 274]). The process of equilibration may be regarded as a succession of steady states with intermediate "transitions through instability", and "the process of equilibration eventually creates the conditions for a new non-equilibrium state" [3, p. 93]. In terms of methodology the comparison of mental and thermodynamic processes seems to be both interesting and promising for the modelling of scientific development (see [19] and [34] for adaptive models of science based on Piaget's theory), although the extent of the analogy should be carefully and critically analysed.

ECOLOGICAL MODELS OF A SCIENTIFIC COMMUNITY

Scientific Community as an Ecosystem

Prigogine's concept of a dissipative structure makes it possible to compare a scientific community to an ecological system and, consequently, apply the mathematical models available in environmental science to the operation and development of science. The methodological aspect of this comparison is

discussed in [33] which establishes a correspondence between the major
properties of an ecosystem and a scientific community. A similar kind of
parallel is drawn between science and the biosphere (a "superecosystem") in
[13] and in [10], where ecological concepts are used for analysing the
relationship between technological projects. The interesting point about
[33] is that it uses the thermodynamic theory of open systems for
mathematical modelling of scientific processes.

In the opinion of the author the chemist Thomas Blackburn, a scientific
community "produces, structures and exchanges information as an ecosystem
produces, structures and exchanges biomass" [33, p. 1141]. Three structural
levels are identified within a scientific community: the "physical"
structure, i.e. structure in the literal sense of the term (institutions,
material conditions, etc.), social structure (teaching staff, professional
communities, specialization, etc.) and the "intangible" intellectual
structure (language, models, logic and other means of organizing knowledge).
The three interrelated structures are basically dissipative, since the
interruption of the fluxes (grants, information, etc.) which feed and
support them results in their destruction (dissipation).

Blackburn interprets an ecosystem as a group of living organisms having a
common access to the source of energy and related to each other by a network
of food and information fluxes. He postulates the following major properties
of such a system:

(a) dependence on the energy and information fluxes, such as solar energy,
    genetic information, etc., which support the system's structure (the
    system decomposes once these fluxes are interrupted);

(b) the ability for homeostasis that ensures the stability of an ecosystem, as
    a rule through negative feedback;

(c) evolution towards greater structural complexity and stability;

(d) limiting factors, i.e. constraints imposed on the rate of mass transfer,
    energy fluxes, etc., which ultimately determine the parameters of an
    ecosystem.

Blackburn points out similar properties in a scientific community. Grants,
scientific equipment and other material resources perform the role of energy
fluxes. These fluxes mainly occur at the level of "physical" structures
(research centres etc.). At the social level the "financial and economic"
fluxes are supplemented by a flux of psychological motivation stemming from
the specific functions performed by various scholars. Finally, at the
intellectual level there is cognitive motivation, i.e. a flux of
"motivational energy" stimulating the search for new experimental data,
problem areas, etc. (see the information flux defined below). Failure of
"motivational energy" arrests the growth of scientific knowledge and results
in a gradual disintegration of the intellectual structure.

The homeostatic properties of a scientific community consist, according to
Blackburn, in regulative factors like historical traditions (system of
education, textbooks, manuals, etc., ensuring the continuity of science),
institutional norms etc., which condition a stable operation of science.

Science evolves through ups and downs of productivity; the beginning of each
cycle is marked by the emergence of a new paradigm (in Kuhn's sense) and the
system's transition to a more "active" (less stable) state. With reference

to Griffith and Mullings [6], who analyse the behaviour of a group facing a new paradigm, Blackburn points out that "communication lines become short and fast; motivational and energy fluxes grow, diversity within an active research group decreases, i.e. all members of the group are absorbed by a new paradigm" [33, p. 1145].

However, this state of affairs cannot last long; the ties weaken, specialization increases and the system is gradually stabilized in its new "normal" state.

The simplest limiting factors in science are financial restrictions, researchers dropping out of the community and other material limitations curbing the growth of a scientific community. More interesting constraints act on the rate of scientific results ("the output of the paradigm") or the rate of information processes. In a biological ecosystem the limited rate of the food (energy) influx causes a greater divergence of the species. As a result the system's diversity increases (the trophic relations become more complex etc.), leading to a higher level of stability, which is ultimately determined by the feasible fluxes of solar energy and food. In the same way the development of a "mature" scientific discipline causes a growing diversity of the scientific community: the emergence of specialization, which is a natural consequence of the limited "throughput" of the community, the appearance of "niches" conforming to the various styles of research (the subdivision of scholars into "artists", "critics", "keepers of records", etc.). It is interesting that there are no such niches in young groups that are still taking shape around a new paradigm.

The natural tendency of a scientific community to fix the existing intellectual and administrative structure, says Blackburn, runs counter to the need for a periodical destruction of these structures by new paradigms for the sake of further development. Therefore managerial strategy should consist in reasonably disturbing the steady "mature" branch with "energy splashes" (provocative questions asked at seminars, violation of boundaries between disciplines, etc.) so as to disrupt its stability and return it to the "active" state for subsequent progress to a higher level.

This is, in a first approximation, Blackburn's ecological approach to a scientific community, which, in our opinion, allows science to be mathematically modelled making use of the similarity between a scientific community and an ecological system. The two models treated below are largely qualitative, for it would be extremely difficult to quantify such major notions as scientific knowledge, the value of a scientific result, problems, objectives and criteria of research, etc. It should be borne in mind, however, that qualitative analysis in mathematics, which is normally based on a qualitative theory of differential equations [1], often answers fundamental questions pertaining to the development of a complex system, its stability, etc. in a general and universal way, which is no less (sometimes more) important than purely quantitative results.

Flow Models and Volterra Equations

By analogy with a biological population regarded as an "elementary evolutionary structure" [26], I shall consider a population of researchers within a scientific community who work in the same problem area on a "group of closely interrelated problems" [9, p. 189]. The concept of problem areas as elementary units of scientific development which emerge, exist and die after being exhausted or ramify and form disciplines, is intensively used by contemporary students of science. Owing to the unity of the subject matter, a problem area is easily (empirically) identified and displays the major numerical trends of science, such as the logistic growth of the number of scholars and publications. A community working in one problem area is limited to 100-200 scholars.* We are interested in the problem area inasmuch as it seems to be an elementary group of scholars having common pursuits and can be regarded as the simplest comparatively homogeneous entity with a single law of operation.

The dynamics of a scientific community working in a given problem area (migration of scholars, which is treated at length in [41]) is composed of two opposed processes: a continuous influx of scholars attracted by prospects of research and a "guarantee" of scientific results and a no less continuous loss of scholars upon the completion of various projects or the "exhaustion" of the problem area. Obviously the relations among scholars form a dissipative structure (in Prigogine's sense) whose stability is maintained by an information flux through the problem area. This flux emerges during the "development" of the problem area and permanently involves new scholars and stimulates further research; it is made up of empirical data (facts, observations, experiments), theoretical results, specific problems, etc. A problem area is therefore a potential source ("intellectual stimulant") of scientific results. Its efficiency may, for instance, be characterized by the amount of information "consumed" by the scholars, the number of scientific problems, etc. (one can also consider the "energy" component of the flux, i.e. grants, equipment, etc.). The exhaustion of a problem area causes the information flux to peter out while the scholars migrate to more promising areas, so that the communication network of the scientific community in question gradually decomposes (possibly until another upsurge or ramification of the problem area).

The operation of a problem area is mathematically modelled in the following way. It is natural to assume that in a stable, steady-state situation information flux I is fed to a community of x scholars at a constant rate v. The incoming information may change (diminish) both because of its fruitful use by the scholars and because of dissipation. The information used by the scholars is transformed (structured) into scientific knowledge, whereas the information that is not used for various reasons (inaccessibility, triviality or overcomplexity, etc.) gets irreversibly dissipated (obsolescence may also be a form of dissipation).

It would be natural to assume that the increment of information due to its use ΔI_1 over time Δt is proportional to the volume of scientific community x with the coefficient of proportionality $V(I)$ depending on the available information: $\Delta I_1 = V(I) x \Delta t$. The loss of information due to dissipation ΔI_2 is directly proportional to the available information I, which ensues from physical considerations: $\Delta I_2 = \alpha I \Delta t$ (α is the coefficient of proportionality). Then the change of information flux ΔI through the area over time Δt is determined by the expression:

$$\Delta I = v \Delta t - \Delta I_1 - \Delta I_2 = v \Delta t - V(I) x \Delta t - \alpha I \Delta t. \qquad (3)$$

---

*Mulkay has used the concept of a problem area as a basis for his ramified model of scientific development [41] which postulates that science evolves through the emergence and disintegration, ramification and intersection of problem areas due to the migration of scholars (forming provisional "social networks") from one problem area to another. For detailed analysis of these processes also see [35].

It may be assumed that the increment of the number of scholars Δx_I over time Δt is proportional to the information used: $\Delta x_I = k\Delta I_I \Delta t$ (k is the coefficient of proportionality). By analogy with the dissipation of information, we shall regard the number of scholars Δx_2 dropping out of the area over time Δt as proportional to the available number of scholars x: Δx: $\Delta x_2 = \beta x \Delta t$ (β is the coefficient of proportionality). To make the result more general, we allow for the additional possibility of scholars entering the area at a constant rate μ as a result of random diffusion, and obtain the following final expression for the increment of the number of scholars Δx over time Δt

$$\Delta x = \mu \Delta t + \Delta x_1 - \Delta x_2 = \mu \Delta t + kV(I)x\Delta t - \beta x \Delta t. \qquad (4)$$

Equations (3) and (4) are divided by Δt, one lets Δt tend to zero, passes to the limit and finally obtains the following system of differential equations which describes the relation between the information flux $I(t)$ and the number of scholars in the community $x(t)$

$$\frac{dI}{dt} = \nu - \alpha I - V(I)x; \quad \frac{dx}{dt} = \mu + kV(I)x - \beta x. \qquad (5)$$

For further analysis the equations are transformed in the following way. Let the rate of the input information flux be $\nu \equiv \alpha I_0$, where I_0 is the input density of the flux and α is its intensity. Now assume (for the sake of simplicity) that the intensities α, β of dissipative processes are identical for information and for the population of scholars: $\alpha = \beta \equiv Q$, and $\mu = Q$, i.e. assume that the influx of scholars into the problem area is determined by the research prospects alone. Furthermore, let the function $V(I)$ be a linear relation, i.e. $V(I) = \lambda I$ (λ is the coefficient of proportionality). It means that the increment of information ΔI_I is proportional to the available information and the population of the scientific community: $\Delta I_I = \lambda I x \Delta t$. Bearing in mind these assumptions we can represent equations (5) as follows:

$$\frac{dI}{dt} = Q(I_0-I) - \lambda Ix; \quad \frac{dx}{dt} = k\lambda Ix - Qx. \qquad (6)$$

It is noteworthy that the same equations describe, *inter alia*, the growth of cell populations, bacterial cultures and other kinds of biomass in a cultivator: the growth of biomass x is a function of the consumption of incoming food substrate I [23, 28]. Generally speaking it is the "absorption" (assimilation, according to Piaget [27]) of the information flux that causes a problem-area community to grow. In other words, the community "digests" scientific information (thereby making further studies possible) and "multiplies", i.e. attracts newcomers by the research prospects. In the same way the loss of scholars may be compared to the death of individuals in a biological population or cells in biomass due to insufficient nourishment.*

Systems described by equations (6) are called flow systems (or systems of flow-type cultivation) because of the steady-state conditions maintained by input and output fluxes. These systems are intensively studied by mathematical biology in terms of their stability, the time dynamics of

---

*One should note that the "epidemic" equations (1) discussed in the first section are a particular case of equations (5) for the linear dependence $V(I) = \lambda I$. In Goffman's "epidemic" model the number of potential researchers S corresponds to information flux I, and the number of active scholars corresponds to community size x.

steady states, etc. [28, 28]. Some results of these studies may be used for mathematical analysis and interpretation of relations (5) and (6) obtained in this paper.

Specifically, (6) is integrable [23]. The solution is interesting for us in that in an asymptotic situation with $t \to \infty$ it yields the following logistic dependence for $x(t)$

$$x(t) = \frac{L}{1 + ae^{-\lambda tL}} \qquad (7)$$

where $a = L/x(0)$ is a coefficient determined by the initial parameters of the scientific community (*inter alia*, the initial "size" of the community $x(0)$); $L = kI_0 - Q/\lambda$ is the so-called "environmental capacity" which determines the ultimate capabilities of the community's steady state; λ is the coefficient of the utilization of information; k is the coefficient of community growth; Q, I_0 are parameters of the information flux (intensity and input density).

Expression (7) means that the time dynamics of the scientific community in a problem area is described by a logistic dependence, which accords with empirical data reported in [35]. This is a convincing piece of evidence in favour of the above model for the dynamics of a scientific area.

A study of the stability of this kind of system yields an interesting qualitative result. One may demonstrate that the system is stable if $dx/dt = k\lambda Ix - Qx \geq 0$, i.e. if $k\lambda I \geq Q$. For the maximum value of the information flux $I_{max} = I_0$ we get $k\lambda I_0 \geq 0$. This is a necessary condition for the system's stability. If the parameters do not meet this condition, for instance if the flux rate Q increases, the steady state is disrupted (stability is disturbed), which, for a biological community, leads to an irreversible deflux ("washing out") of the biomass from the cultivator. In our case it corresponds to a decrease of the input information density $k\lambda I_0 < Q$ due to the exhaustion of the problem area (specifically, the depleted problem-solving capability of the existing paradigm). An irreversible process sets in: scholars leave the area without replacement and migrate to a more promising field. This is empirically noted and studied by science sociologists [35, 41] as the process of area decomposition, differentiation, ramification, temporary "damping", etc.

Note that the presence of input and output fluxes in "flowtype" models (5) and (6) for the operation of a problem area is both an advantage (completeness) and a disadvantage (complexity). To simplify the model one may assume that $Q = 0$ in (6); this would mean the absence of fluxes, i.e. would make the system closed. The dependence is thereby rendered less general but may be more thoroughly analysed, which is important for relatively simple situations. We shall consider the representation of a "closed" problem area with an *a priori* fixed set of unsolved problems that are being successively solved by the scientific community, the results of the solution appearing in the form of publications.

It seems natural that the probability of publishing another paper (solving another problem) will be determined both by the number of papers already published y (facilitating the solution of the new problem) and the possibility of finding an as yet unsolved problem, which depends on the available number I of unsolved problems as a function $V(I)$. Hence it is assumed that the increment of the number of publications Δy over time Δt is proportional to the product of $V(I)$ and y with a proportionality coefficient k: $\Delta y = kV(I)y\Delta t$. Similar considerations, taking into account the closed nature of the problem area, lead to the following expression for the decrease in the number of unsolved problems ΔI over time Δt: $\Delta I = V(I)y\Delta t$. Hence we obtain a system of equations characterizing the "transformation" of problems into publications with a coefficient k

$$\frac{dI}{dt} = -V(I)y; \quad \frac{dy}{dt} = kV(I)y. \qquad (8)$$

Assuming the function $V(I)$ to be linear: $V(I) = \lambda I$ (i.e. supposing that the possibility of finding an unsolved problem is proportional to the number of problems with a coefficient λ) we arrive at (6) for the case $Q = 0$, as might have been expected (there is no flow)

$$\frac{dI}{dt} = -\lambda Iy; \quad \frac{dy}{dt} = k\lambda Iy. \qquad (9)$$

System (9) has an integral (Monod integral): $y + KI = \text{const} \equiv P$, while $P = y(0) + kI(0)$, where $I(0)$, $y(0)$ are the initial values of the set of problems and the set of publications respectively. The Monod integral leads to a logistic equation for the publication dynamics, whose solution is described by the logistic dependence

$$\frac{dy}{dt} = \lambda y(P-y); \quad y(t) = \frac{P}{1 + be^{-\lambda tP}} \qquad (10)$$

where b is a coefficient determined by the initial conditions.

A logistic dependence for the publication dynamics describing the progressive depletion of a problem area is noted in many studies, e.g. for the dynamics of publications on a specific psychological problem or in the field of the games theory [43], on rural sociology or on the theory of finite groups [35], etc. These empirical data support the proposed model for the operation of a problem area (9).

The relationship between the dynamics of a community of scholars $x(t)$ and the array of their publications $y(t)$ may be studied by means of the following model. According to Mulkay [41], new scholars enter an area mostly as a result of contacts with those who are already engaged in it. It is therefore natural to assume that the increment of the number of scholars Δx_1 over time Δt is proportional to the size of the existing community: $\Delta x_1 = a_1 x \Delta t$ (a_1 is the coefficient of proportionality). We assume that each scholar who has published a certain number of papers leaves the area, i.e. that the loss of scholars from the area Δx_2 over time Δt is proportional to the product of the number of scholars and the number of publications: $\Delta x_2 = b_1 xy \Delta t$ (b_1 is the coefficient of proportionality). The net increment of the community over time Δt equals the difference between the influx Δx_1 and the deflux Δx_2

$$\Delta x = \Delta x_1 - \Delta x_2 = a_1 x \Delta t - b_1 xy \Delta t. \qquad (11)$$

To determine the increment of the array of publications we assume that all scholars may publish papers with equal probabilities and these probabilities depend on the total number of papers published by the community, indeed are proportional to it in the simplest case. Therefore the increment of publications Δy_1 over time Δt is proportional to the product of the number of scholars and the number of papers published: $\Delta y_1 = a_2 xy \Delta t$ (a_2 is the coefficient of proportionality). The area loses publications as a result of obsolescence, which is proportional to the available number of publications. Therefore over time Δt the active array of papers decreases by

$\Delta y_2 : y_2 = b_2 y \Delta t$ (b_2 is the coefficient of proportionality). The net increment of the number of publications over time Δt equals the difference between the number of papers published Δy_1 and that of obsolete publications Δy_2

$$\Delta y = \Delta y_1 - \Delta y_2 = a_2 xy \Delta t - b_2 y \Delta t. \tag{12}$$

Now equations (11) and (12) are divided by Δt, one passes to the limit and obtains systems of differential equations for the relation between the growth of the community and the number of publications

$$\frac{dx}{dt} = a_1 x - b_1 xy; \quad \frac{dy}{dt} = a_2 xy - b_2 y. \tag{13}$$

This system of Volterra equations is well known in mathematical ecology [4]: *inter alia*, it describes an interspecies relationship of the "predator-victim" type [4]. When interpreted in terms of scientific development, these equations and their analysis [25] may be used for simultaneous consideration of the scientific community x and its output y.

For example, by setting dx/dt and dy/dt in (13) to zero we obtain steady-state solutions of the system: $x_0 = b_2/a_2$, $y_0 = a_1/b_1$, which depend on the coefficients of increment a_1, a_2 and loss b_1, b_2. It is an interesting feature of processes described by Volterra equations that they produce fluctuations when the initial states $x(0)$, $y(0)$ deviate from steady states x_0, y_0. Studies of scientific development have indeed registered such fluctuations, which gives more credit to model (13). For instance, Serov [24] reports fluctuations in the publication activity as related to a scholar's range of interests, fluctuations in the current evaluation of scientific texts by the scientific environment, in the intensity of information consumption in libraries, etc. Rainoff [46] investigates the fluctuating growth of the number of discoveries in physics in various countries, Ayres [31] dwells on fluctuations in various branches of science and technology. A model based on Volterra equations (13) may be used for a specific study of such "auto-oscillatory" phenomena in science.

For example, Volterra equations are used by Müller for analysing the relationship between problems x and knowledge y emerging from the solution of the problems. The equations themselves (see [42], a concise outline in [12] are based on the following assumptions: the increment of knowledge and the reduction of problems are proportional to their product xy; if there is no scientific activity, problems grow in proportion to their number $a_1 x$ (since no problems are solved), while knowledge diminishes in proportion to the available knowledge $b_2 y$ (becomes obsolete). The author compares the quantitative solution of the resulting system [one like (13)] with empirical measurements (knowledge is identified with the number of publications) and notes good agreement between the two.

Competition and Differentiation Models

When the research possibilities of a problem area are exhausted, its further destiny (termination, differentiation, change of paradigm, etc.) depends on the choice of one of the many competing hypotheses. We shall now proceed to study the competition of hypotheses by means of a comparatively simple mathematical model which is normally used for ecological analysis of the competition of biological species, genotypes, etc. [22].

Suppose there are two scientific theories both claiming to offer a satisfactory explanation of the same empirical data and therefore competing with each other. To characterize these theories we shall consider the changing sum of information $I_1(t)$, $I_2(t)$ in favour of theory T_1 or theory T_2, respectively. This sum may include facts and observations, the number of publications in support of the first or the second theory, the number of scholars adhering to T_1 or T_2, etc. We are interested in the qualitative result of "survival" of one of the theories or their "coexistence", so that the numeric value of I_1, I_2 is immaterial [unlike, say, $I(t)$ in (5)].

Suppose that in the absence of the second theory T_2 the information flux I_1 corresponding to theory T_1 is described by a common logistic dependence: $I_1 = I_1(a_1 - b_1 I_1)$, where a_1, b_1 are coefficients that reflect, respectively the dynamics of information growth with an intensity a_1 and the dynamics of data depletion (obsolescence) with an intensity b_1. In the presence of a competitive theory T_2 one should, apart from the factors intrinsic to T_1, take into account the external factor of T_2's inhibiting effect on T_1, e.g. the results obtained by T_2 adherents (information I_2) that throw doubt upon T_1 ("adulteration" of T_1). We assume that this effect is proportional to the product $I_1 I_2$, i.e. that it decreases the rate of flux I_1 by the value $c_1 I_1 I_2$ (c_1 is the coefficient of proportionality). Since the same holds for the flux I_2 corresponding to theory T_2, we obtain a system of equations that describes the competitive interaction of theories T_1, T_2 through their respective information fluxes (the meaning of coefficients a_2, b_2, c_2 is analogous to that of a_1, b_1, c_1)

$$\frac{dI_1}{dt} = a_1 I_1 - b_1 I_1^2 - c_1 I_1 I_2;$$

$$\frac{dI_2}{dt} = a_2 I_2 - b_2 I_2^2 - c_2 I_1 I_2. \qquad (14)$$

The system of equations (14) (as well as its n-dimensional variant

$$\frac{dI_i}{dt} = \left(a_i - \sum_{j=1}^{n} \gamma_{ij} I_j \right) I_i$$

where $i = 1, 2, \ldots, n$; $a_i > 0$ is a growth coefficient; γ_{ij} is the coefficient of environmental influence) is analysed in many studies dealing with the competition of species in a biological community [22, 25]. These studies show that the stability of (14) is determined by the relation between the product of flux reduction coefficients b_1, b_2 and the product of inhibition coefficients c_1, c_2. If $b_1 b_2 > c_1 c_2$, i.e. the "intrinsic" factors have greater import for the development of a species than competition, then system (14) is stable: the species may coexist with a constant relationship between them. If, however, $b_1 b_2 < c_1 c_2$, i.e. interspecific antagonisms play the key role, then the system is unstable: only one of the competing species will survive, depending on the initial conditions. This conclusion, which is known in ecology as Gause's principle (the principle of competitive elimination), means that two different species cannot exist in one ecological niche (it has been empirically tested on many occasions by placing two closely related species into one habitat [18, 28]).

We proceed to generalize these results onto the interaction of information fluxes I_1, I_2 for competitive theories T_1, T_2. Obviously when there is little evidence within T_2 refuting T_1 (or vice versa), i.e. when the theory evolves on the basis of its intrinsic mechanisms rather than competition ($b_1 b_2 > c_1 c_2$), two competing theories may steadily coexist. In the same way competing scientific schools may coexist when either lacks conclusive arguments to "adulterate" the competitor's theory. For instance, with regard to the competing theories of Kuhn and Popper it is stated in [36, pp. 49-50] that the history of scientific development is not yet sufficiently studied to allow a choice between the two approaches. In the opposite case, when the competition between theories is keen ($b_1 c_2 < c_1 c_2$), only one theory "survives", i.e. the evolution of science follows a pattern similar to Gause's competitive elimination. Analysis of the competition of theories in a scientific community may prompt the subsequent trend of the research process for a given choice of its parameters.

This simplest model of theoretical competition is phenomenological since it does not explicity allow for the part played by the community of scholars in the interaction of scientific theories. The inclusion of this factor leads to a more complicated model.

Let y_1, y_2 be the number of publications by scholars adhering to theories T_1 and T_2, respectively, while x is the population of the community. The productivity of either theory depends on the number of scholars x and is determined by the function $V_i(x)$ (i=1,2). Scholars enter the problem area (the area of competition between T_1 and T_2) at a constant rate v and leave the area either having publications in support of theories T_1, T_2 (in this case the dimunution of the community is proportional, respectively, to products $V_1(x)y_1$ or $V_2(x)y_2$) or having no such publications, e.g. because of incorrect choice of problems (the diminution of the community is proportional to the available number of scholars x). Fluxes y_1, y_2 grow owing to the increment of publications in proportion to products $V_1(x)y_1$, $V_2(x)y_2$ and decrease due to the obsolescence of publications in proportion to their available number y_1, y_2. Proceeding from these assumptions and taking $V_i(x) = a_i x/b_1 + x$, i.e. a saturation function, we get the following system of equations (α_1, α_2 are coefficients of proportionality, Q is the coefficient of dissipation, $v \equiv Q x_0$)

$$\frac{dx}{dt} = -\alpha_1 \frac{a_1 x}{b_1 + x} y_1 - \alpha_2 \frac{a_2 x}{b_2 + x} y_2 + Q(x_0 - x);$$

$$\frac{dy_1}{dt} = \frac{a_1 x}{b_1 + x} y_1 - Q y_1; \quad \frac{dy_2}{dt} = \frac{a_2 x}{b_2 + x} y_2 - Q y_2 \tag{15}$$

Analysis of this system of equations [21, 28] shows that no two theories can coexist for a long time within one scientific community (which is a limiting factor here): the more constructive theory eventually ousts the other, thereby determining the further evolution of science.

In conclusion one should note that limiting factors in general are one of the basic conditions for the development of science (the emergence of a new theory, differentiation of research areas, etc.). Limitations (with respect to material resources, information fluxes and other parameters conditioning the operation of a scientific community) are a prerequisite for the choice of a better theory. Otherwise, instead of goal-oriented development, science would be subject to a divergent process involving a multiplicity of coexisting scientific explanations. The limiting factor in ecology is, for instance, the food substrate, whose depletion compels a species to look for new sources of nourishment, leading to the fixation of new functions and, eventually, to the evolution of the biological community and the differentiation of species [21]. In the domain of science the depletion of a problem area stimulates the migration of scholars to another area or the ramification of the area concerned. However, these are not the only factors that condition the evolution and differentiation of science. As shown in

[22, 23], within a mathematical model of the competition of species in a biological community the diversity of the community (i.e. the number of species) should grow with its size in the course of evolution. The more complex is the community, the more extensive and diverse are the trophic relations among species etc., the more stable is the biological community in question [22, 23]. Similar models may be used for analysing the differentiation of science in the course of development. One may suppose that as the "mass" of scientific knowledge grows, its differentiation into disciplines, problem areas, etc. should also grow, and that the process of differentiation is inseparable from an inverse process of integration, i.e. the formation of increasingly complex bonds among the various problem areas.

CONCLUSION

The development of any complex system involves two associated processes: maintenance of stability and temporary disturbance. The stability of a system means its ability to resist perturbation: when its steady state is disturbed, a stable system regains it by means of compensatory action. This is a necessary element in the operation of complex systems, for the maintenance of stability ensures the system's "survival" and the continuity of its development, and most important of all, the preservation of "old" features ensures the assimilation of "new" ones.

The instability emerging in the course of evolution allows the system to leap to a new state.

Generally speaking, this leap may be regarded as the system's reaction to, and compensation for perturbation. The system assumes a new state instead of reverting to the old one, i.e. "development through instabilty" ensures stability at a higher level. Now stability itself is understood not as the static stability of equilibrium structures like crystals but as a dynamic stability of open systems based on self-organization and self-regulation which complex systems mainly achieve through information exchange with the environment (dissipative structures [5] or dynamic regulators [14]).

With regard to scientific systems, one seems justified in claiming that their stability in the steady state of "normal science" and instability at the time of "scientific revolutions" are equally indispensable to development. The solution of problems which depletes the existing paradigm and the generation of new problems menacing the paradigm end up by driving the paradigm (and the scientific community) from stability to instability; the emergence of a fundamentally new, usually unpredictable, scientific result (creative "fluctuation") results in a spasmodic change of the paradigm.

This process is fairly well described by the thermodynamic theory of open systems, which postulates that the hierarchic organization of a developing object is the result of an evolutionary succession of increasingly complex structures. By relating "the existence of different levels of organization to the sequence of instabilities" the theory shows that "a state of a given complexity may possess a "memory" of the past instabilities each of which may contribute to the emergence of a new feature that is essential to stability and maintenance of the final state" [17, p. 522].

In other words, knowledge is not "drawn" out of the environment and added to the past knowledge; it is created as a result of scientific development: unorganized information is converted into organized knowledge by the scientific community, which acts as a "nonlinear converter". We have taken three parameters (information flux measured, for instance, by the number of input problems I; size of the scientific community determined by the number of scholars x; output knowledge measured by the number of publications y) to be dynamic variables related as I→x→y. The systems of equations which establish relations between these variables: I-x, I-y, x-y are based on the following assumptions: the rate of change is proportional to the magnitude of the variables, i.e. the "birth" and "death" of the elements is proportional to their available number in the system; the interaction of the variables is proportional to the product of their magnitudes, i.e. the product of the sizes (numbers of elements) of the interacting systems. These assumptions are widely and successfully used in various domains, from the kinetics of physico-chemical reactions through the interaction of biological populations and socioeconomic systems; they are simple enough to allow mathematical modelling of the system concerned. A number of conclusions suggested by the mathematical modelling of a scientific community have been confirmed by empirical measurements or accord with qualitative concepts concerning the behaviour of the systems in question.

It should be noted that the models discussed in this chapter have suggested several hypotheses on the development of science (e.g. the hypothesis about the instability of scientific systems during the transition to a new paradigm etc,) that have been empirically tested. Besides, the various models may be used for practical studies (and utilization) of the formation of scientific information arrays, for analysis of the dynamic interaction of an information array and the scientific community, for scientific and technical forecasting, etc. Of course these are preliminary models designed to illustrate the applicability to science of the mathematical tools proposed for other complex systems (e.g. ecological ones). They only reflect the most obvious features of scientific development, raise more problems than solve and, naturally, require a thorough analysis of the methodological premises, and more empirical measurements, say a quantification of the parameters involved. Nevertheless, these models may serve as one of the starting points for further research in the field of mathematical simulation of science.

REFERENCES

1. A. A. Andronov, E. A. Leontovich, I. I. Gordon and A. G. Mayer, *Qualitative Theory of Second-order Dynamic Systems*, Moscow (1966), in Russian.
2. T. J. Bailey Norman, *The Mathematical Approach to Biology and Medicine*. London, John Wiley (1967).
3. T. W. Barrett, Entropy and symmetry - their relation to thought processes in the biological system, *Kybernetik*, 6, 102-112 (1969).
4. Vito Volterra, *Leçons sur la theorie mathématique de la lutte pour la vie*. Paris, Gauthier-Villars (1931).
5. P. Glansdorf and I. Prigogine, *Thermodynamic Theory of Structure Stability and Fluctuations*. Wiley-Interscience (1971).
6. B. C. Griffiths and N. C. Mullins, Coherent social groups in scientific change ("invisible colleges" may be consistent throughout science), *Science* 177 959-966 (1972).
7. V. V. Ivanov, *Essays on the History of Semiotics in the USSR*. Moscow (1976), in Russian.
8. *The Conception of Science in Bourgeois Philosophy and Sociology*. Moscow (1976), in Russian.
9. D. Crane, Social structure in a group of scientists: a test of the "invisible college" hypothesis, *American Sociological Review* 34, 335-352 (1969).

10. B. I. Kudrin, Application of biological concepts to the description and prediction of large-scale technologically formed systems, in: *Electrification in Iron and Steel Industry in Siberia*, issue 3, Tomsk (1976), in Russian.
11. Th. S. Kuhn, *The Structure of Scientific Revolutions*. Chicago, University of Chicago Press (1962).
12. F. Müller, On one hypothesis of scientific progress, *Naukovedeniye i Informatika*. issue 15. (1976), in Russian.
13. V. V. Nalimov and Z. M. Mulchenko, Science and biosphere, *Priroda* No. 11 (1970), in Russian.
14. L. A. Nikolaev, *The Fundamentals of Physical Chemistry of Biological Processes*. Moscow (1976), in Russian.
15. J. Piaget, *La psychologie de l'intelligence*. Paris, Collin (1967).
16. I. Prigogine, *Introduction to Thermodynamics of Irreversible Processes*. Springfield (1955).
17. I. Prigogine and G. Nicolis, Biological order, structure and instabilities, *Quarterly Review of Biophysics* $\underline{4}$, 107 (1971).
18. A. Rapoport, Uses of mathematical isomorphism in general system theory, in: G. J. Klir (ed), *Trends in General Systems Theory*, pp. 42-77. New York, Wiley-Interscience (1972).
19. L. A. Rastrigin and V. A. Markov, *Cybernetic Models of Cognition*. Riga (1976), in Russian.
20. N. I. Rodny, *Essays on History and Methodology of Natural Science*. Moscow (1975), in Russian.
21. Yu. M. Romanovsky, N. V. Stepanova and D. S. Chernavsky, *Mathematical Modelling in Biophysics*. Moscow (1975), in Russian.
22. Yu. M. Svirezhev and E. Yu. Elisarov, *Mathematical Modelling of Biological Systems*. Moscow (1972), in Russian.
23. Yu. M. Svirezhev, On mathematical models of biological communities and related management and optimization problems, in: *Mathematical Modelling in Biology. Proceedings of 1st School of Mathematical Modelling of Complex Biological Systems*. Moscow (1975), in Russian.
24. P. K. Serov, On the rhythms of processes in scientology, in: *Science and Technology (History and Theory)*, issue 6. Leningrad (1971), in Russian.
25. John M. Smith, *Models in Ecology*. Cambridge, Cambridge University Press (1971).
26. N. V. Tymofeyev-Resovsky, N. N. Vorontsov and A. V. Yablokov, *A Brief Review of Evolution Theory*. Moscow (1969), in Russian.
27. J. H. Flavell, *The Developmental Psychology of Jean Piaget*, Princeton, New Jersey, van Nostrand, (1962).
28. I. A. Shytov, Mathematical models of cellular population growth, in: *Mathematical Modelling of Complex Biological Systems*. Moscow (1975), in Russian.
29. M. Eigen, Molecular self-organization and the early stages of evolution, *Quarterly Review of Biophysics* $\underline{4}$, 149 (1971).
30. M. Eigen, Self-organization of matter and the evolution of biological macromolecules, *Die Natirwissenschaften* $\underline{58}$ (1971).
31. R. U. Ayres, *Technological Forecasting and Long-range Planning*. New York, McGraw-Hill (1969).
32. A. I. Yablonsky, On extremal features of random search, *Izv. Vuzov. Radiofizika* $\underline{XIV}$, No. 7 (1971), in Russian.
33. T. R. Blackburn, Information and the ecology of scholars. *Science* $\underline{181}$, 1141 (1973).
34. R. J. Blackwell, The adaptation theory of science, *International Philosophical Quarterly* $\underline{8}$, No. 3 (1973).
35. D. Crane, *Invisible Colleges. Diffusion of Knowledge in Scientific Communities*. Chicago, University of Chicago Press (1972).

36. Criticism and the growth of knowledge. *Proceedings of the International Colloquium in the Philosophy of Science*, Vol. 4, Cambridge (1970).
37. Goffman W. A general theory of communication, in: *Introduction to Information Science*. New York (1970).
38. W. Goffman and G. Harmon, Mathematical approach to the prediction of scientific discovery, *Nature* 229, No. 5280 (1971).
39. R. L. Hamblin, R. B. Jacobsen and J. L. Miller, *A Mathematical Theory of Social Change*. New York, John Wiley (1973).
40. M. B. Line and A. Sandison, "Obsolescence" and changes in the use of literature with time. *Journal of Documentation* 30, 283 (1974).
41. M. J. Mulkay, Three models of scientific development. *The Sociological Review* 23, 509 (1975).
42. F. Müller, Ein Versuch zur warscheinlichkeitstheoretischen Erklärung der wissenschaftlichen Produktivität. *Elektronische Informationsverarbeitung und Kybernetik* 10, No. 1 (1974).
43. M. Nowakowska, *Psychologia ilościowa z elementami naukometrii*. (Warszawa) 1975.
44. J. Piaget, La psychologie, in: *Tendences principales de la recherche dans les sciences sociales et humaines*. Paris (1970).
45. I. Prigogine, Time, irreversibility and structure, in: *The Physicist's Conception of Nature*. Dordrecht (1973).
46. T. J. Rainoff, Wave-like fluctuations of creative productivity in the development of West European physics in the XVII and XIX centuries, *ISIS* XII, No. 38 (1929).
47. S. Toulmin, *Human Understanding*, Vol. 1. Oxford (1972).

DISCIPLINARY KNOWLEDGE AND SCIENTIFIC COMMUNICATIONS

A. P. Ogurtsov

Scientological studies successfully apply methods of epistemology, informatics, sociology, etc. Their use involves the adoption of certain units of analysis and specific dichotomies on the basis of which they evolve in the relevant scientific disciplines. Thus epistemological methods of studying scientific knowledge are based not only on a specific concept of science which predetermines certain dichotomies of knowledge (theoretical and empirical levels, theory and method, etc.) but also on an abstraction from the specificity of disciplinary knowledge. The theory of knowledge reveals some universals of scientific knowledge, operating regardless of its segmentation into a number of scientific disciplines. As a result, the latter may be described in terms of universal structures, variously defined in different conceptual frameworks (e.g. Kant's transcendental unity of apperception, Fichte's absolute contemplation, Hegel's absolute spirit, etc.).

A historian of science, is confronted primarily with the disciplinary variety of scientific knowledge, and it is this empirical fact that underlies his work. Scientology, oriented towards history of science, cannot, of course, lose sight of some universals of scientific knowledge but, on the other hand it cannot confine itself to them.

There are roughly two trends in the studies of science. One of them is concerned with a search for invariant structures of scientific knowlege and with the isolation of logical-methodological norms and regulatives, operating independently of cognitive acts and communication of scientists. Such studies are dominated by a typological way of thinking, aimed essentially at identifying a single structure, an archetype and a norm and interpreting variability as a secondary and non-essential phenomenon, an indicator of inadequate knowledge of the primordial archetype. Orientation towards invariant structures is deterministic-systemic; history of science is treated irrespective of the activities of scientists, disregarding the fact that the knowledge, required in the system, does not come into being until the scientific community takes notice of it and accepts it.

The other trend is associated with an analysis of acts of research and communication of scientists. Its major goal is a study of variation of science phenomena, the variability of science, various groups of scientists and the scientific community as a whole. In terms of this populationist-systems approach science is interpreted as a set of cognitive acts, occurring in various historical situations. A turn from former methods of describing scientific knowledge (primarily as a system of norms, self-identical structures, language universals, etc.) to a situational analysis of science as research and a communicative process makes it possible to bring out its new dimensions. This 'stratum' of science has gained prominence in the so-called logic of communications, developed by Lorenzen, in hermeneutic science-study procedures, proposed by Gadamer and in the treatment of science as a process of research, an idea, maintained by Thörnebom and Radnizky. Some foreign science methodologists, however, while emphasizing the need of describing specific historical situations in their development, rule out altogether the possibility of defining any universal methodological norms, characterizing scientific knowledge in its entirety. This is essentially the position of the "new left" in science methodology - Feyerabend, Spinner and others [29]. While stressing the communicative aspect of research and its situational determinateness, they actually reduce research to a flow of activity, not subject to any normative-disciplining regulation. Disregard for the need for science to elaborate systematically upon and explain its results and to embody them in objectively meaningful structures, results in giving up altogether the idea of isolating any logical-methodological universals and norms. While emphasizing varietiability of research, they lose sight of the invariant structures of scientific knowledge, including disciplinary knowledge. Finally, the treatment of scientific knowledge as a set of free communicative events denies the existence of the invariants that discipline each communicant thereby restricting his freedom and form the basis of their mutual understanding and implementation of both communicative events and research.

The main challenge to scientologists today is to find ways of integrating invariant structures and variability of scientific knowledge without rendering absolute any dimension of scientific knowledge. In this paper I shall address semiotic methods which, in my view, make it possible to identify in scientific knowledge characteristics, specific to the language of science.

A SEMIOTIC APPROACH TO DISCIPLINARY KNOWLEDGE

Recently semiotics has become one of the most effective instruments of exploring different phenomena of culture (cf. the studies of the Tartu school, the papers of M. Lotman, Uspensky, Ivanov, Toporov and others). A semiotic approach to science involves its treatment as a communicative sign system with at least two aspects — the signs and their meaning. But this is not an end in itself. This approach offers insights into the logical-methodological structure of science with all its differentiation and stratification. The language of science should be viewed not as a single abstract system of self-identical norms (similar to Leibnitz's universal grammar) but as a superposition of disciplinary languages, their norms and patterns. A universal normative-value structure of science which classical methodology sought to define is a potential characteristic actualized in polyglot disciplinary languages and the language of natural scientific communication. Each disciplinary language is specific not only in its lexicon but also in its referential-semantic and normative-value spectrum. It expresses a certain world view, common to all participants of research, and a unique totality of methodological rules and procedures, recognized by the scientific community as logical-methodological norms. Stratification is inherent in the languages of individual sciences as well as in scientific knowledge as a whole. Disciplinary languages of natural sciences may be

...tterns of cognitive acts. In relation to a natural flow of research acts ... is a system of language and norms, the prerequisite for reaching mutual ...derstanding among scientists and implementing cognitive acts. This ...nguage is also the frame of reference and the model of the world that ...rmeate each communicative event. It incorporates the invariant structures ...ognitive, linguistic, normative, etc.) which alone may serve as a basis for ...e diverse forms of communication among scientists. Disciplinary language ... not external to or compulsory for communicative events. It is both a ...ndition and an instrument for a free exploratory search which they need in ...der to achieve their goal - to cognize the reality and express it in a ...rtain universally meaningful language.

...ere are two ways of presenting the interrelationship between disciplinary ...nguage and real-life research and communicative acts. In a sense, this ...nguage precedes communicative events and is a prerequisite for the ...plementation of research. But it is equally true that research acts ...recede the embodiment of the results achieved in the disciplinary structures ...f scientific knowledge.

...e understanding is gaining ground now that not only a scientific discipline ... a system of functionally meaningful norms but that the communicative ...ature of science is its dominant feature.

...t is increasingly realized that disciplinary norms and patterns are not self-...ontained and self-sufficient clichés and stereotypes or invariable and ...etrified clusters of dogmas but relatively invariant ideal models on which ... scientist can base his innovations. The interiorization of these models by ... researcher determines his frame of reference, categorial dichotomies and ...ethodological guidelines which underlie the specific character of his ...pproach to the subject at hand.

...either the progress nor the results nor the subjects of cognition can be ...ivorced from the communicative situation where research takes place. Each ...lement of a cognitive act and of its content is imbued and illuminated with ...he context of communicative interaction. In other words, a semiotic ...pproach provides insights into a new dimension of science - communication ...f scientists - and thereby helps to describe the determination of cognitive ...cts by the communicative context, the mutual orientation of each ...ommunicant, intended meanings behind each communicative event and appeal ...o another equal mind.

...cience turns out to be a two-way contradictory process. It is both a ...otality of objectivated linguistic texts and a set of acts of communication. ...his approach isolates different types of communicative relations: some are ...oncerned with the setting up of norms in research, others with the "ejection" ...f standards and regulatives as patterns and recognized disciplinary norms and ...till others with problems of cooperation among scientists. Meanwhile the ...leading edge" structure in science is interpreted as the coexistence and ...onfrontation of different meanings, as polyglotism of former and future, ...ealized and unrealized meanings and as a universe of intradisciplinary and ...nterdisciplinary languages, reflected in one another. The field of ongoing ...esearch is not only a mixed context of differentiated scientific languages ...nd those in the process of differentiation but also a field of possible ...anguages and their meanings. This "layer" of science, like science as a ...hole, is a multitude of natural dialogues - with both one's contemporaries ...nd one's predecessors. Scientific knowledge turns out to be a dramatic ...nterplay of different meaningful acts, an overlap of semantizations, ...odifications and negations of "already established" meanings. According to ...akhtin the gist of a dialogue is a dramatic confrontation of different

Disciplinary Knowledge and Scientific Communications

viewed as specific strata or levels of scientific knowledge, super
one another from "object" languages, related to the external reali
procedures of empirical observation and measurement, to metalangua
of them has its own specific semiotic-communicative resources, its
semantics and syntax.

Usually a semiotic approach to the language of science is limited
differentiation between its three aspects (syntactic, semantic and
and a rigorous separation of syntax from semantics [15, pp. 11-12]
differentiation and emphasis on logical relations characterize jus
disciplinary language - the language of mathematics where, accordi
Bourbaki, mathematical objects are not so important as their relat
ordering, topological and algebraic structures). The language of
a more complex entity where measurement procedures and quantitativ
[9, pp. 95-179] form the primary basic language, permit the semant
extrasemantic elements and where there is no rigorous distinction
syntax and semantics. The reduction of the entire semiotic approa
differentiation of the three above aspects is based on a methodolo
unjustified extension of the explanatory patterns, developed in st
language of mathematics, to the entire disciplinary corpus of sci
knowledge.

A semiotic approach to disciplinary knowledge makes it possible to
it into two hypostases one of which characterizes it as a system o
the other as a totality of communicative events. Just as modern l
has distinguished since de Saussure between *langue* and *parole*, so
advisable for scientological analysis to draw a similar distinctio
a system of scientific knowledge with its internal determinateness
normative character and an ensemble of scientific communicative ev
other words, the basic opposition is the one between a disciplinar
and scientific communicative events.

The analogy between disciplinary and natural language is wellfound
functions of the disciplinary body of knowledge relative to cognit
communicative events are similar to those of language in relation
activity. The disciplinary structuring of knowledge assigns to th
scientific community a certain system of a universally recognized
a system of norms and patterns for the solution of scientific prob

Disciplinary knowledge as a linguistic modelling semiotic system p
be also a model of the world as perceived by a given scientific co
comprising semiotic systems of different levels. This model of th
essentially determines a mode of action for each member of the sci
community, determining a set of operations, the rules for their us
totality of principles and explanatory patterns, methodological re
etc.

The language of disciplinary knowledge and communicative events di
mode of their existence. Without discussing in depth the ontologic
of disciplinary semiotic structures, merely note that they are a so
product, a totality of language texts, adopted by a given scientifi
and performing a certain social function. The latter is a realizat
each researcher's cognitive abilities and skills and the attainment
understanding among scientists.

Research acts are always closely linked to communication between sc
and are diverse and heterogeneous. They are related to both social
individual realms. Disciplinary language can exist as long as it i
language of scientific communication which determines the technique

meanings [10, p. 300]. A meaning, semanticized in the activity of each participant in the dialogue, borders on other meanings, and the mutual understanding, reached in a dialogue, is localized in the same historical plane. Dialogue is by definition a perpetually changing historical reality, incompatible with a didactic-monologic statement of a single position.

The content and norms of disciplinary knowledge constitute the functionally significant component of the development of science, known as tradition. The mechanisms of tradition (*traditium, tradendum* - something handed down and something being handed down) achieves continuity in the development of knowledge and harmony between those who produced the results, serving as patterns, and those who absorbed them and reveal their new aspects and layers of meaning, modify them and, at the same time, achieve new results. There can be no thinking or creativity outside traditional norms and concepts because tradition reveals historical relation to the past. While stressing the tremendous role of disciplinary traditions in the functioning of scientific knowledge, one should not blindly and unreasonably cling to everything created by past generations or protect outdated traditions from the impact of scientific thinking [1, Vol. 20, p. 455]. It is necessary to bring out the relation between the components of knowledge available and those selected by the researcher and to define the historical prerequisites of cognitive acts. Orientation towards isolation from all traditions makes cognition a series of acts, void of any objective meaning, and reduces tradition from an indispensable element of culture to the opposite of renovation and a set of dogmas, taken for granted.

Disciplinary knowledge preserves the patterns, inherited from the past and handed down to the next generations, making possible not only the continuation of cognition and the reproduction of scientific results but also their effective use and transformation.*

DEVELOPMENT OF KNOWLEDGE OBJECTIVATION FORMS

The approach to science as a semiotic system makes it possible to view it as a development of knowledge objectivation forms, as an accumulation of language types and as a transition from oral speech to printed text. The development of speech forms and language texts cannot be described as the total replacement of the previous form by the subsequent one. It is rather an accumulation of forms where the previous phenomenon and existential pattern of culture coexist with the newly formed phenomena, constituting their source and nucleus [23, p. 323]. Stratification is inherent not only in language and texts. It is a more universal feature of the development of culture and characterizes also the forms of the social and internal structuring of knowledge.

---

*In his essay "On tradition and creative personality" (1919) T. S. Eliot notes a negative attitude to tradition in early twentieth century Anglo-Saxon culture. He adds that tradition implies a sense of history while a sense of history implies the perception of the past not only as the past but also as the present; a sense of history requires that man should write not only as if he carried inside all his generation but also as if all European literature from Homer as well as the entire literature of his own country existed synchronously and were synchronously ordered. It is this sense of history, i.e. an awareness of the extratemporal and temporal, one inside the other and beside the other, that makes the writer traditional and keenly aware of his place in time and of his belonging to his time [26, pp. 95-96].

Historically the initial form of the social transmission of knowledge was the teacher's oral speech and teacher-to-pupil communication in the form of oral speech events.*

This type of language presupposes contact between the communicants, its maintenance and a specific type of utterances by the speaker and the listener. In other words, this type of communication implies: (a) cognitive and verbal competence of each participant, (b) a common intersubjective part in the process of communication to ensure their mutual understanding, and finally (c) new cognitive and linguistic results, produced in the course and as a result of communicative events. Needless to say, this is merely a certain dialogue scheme making it possible to isolate a single "dialogue thread" in communicative events. More complex in its structure is a situation where the speaker is not only a speaker but, at the same time, a listener, and vice versa, i.e. where there is a continuous interchange of social roles and functions of each of the equal partners in communication.

The Socratic dialogues of Plato are a historical proof of the fact that in the early period of the history of culture and education the transmission of knowledge assumed precisely this form. They provide excellent material for a study of specific generative rules for such texts and normative structures organizing the utterances of each side in a certain sequence and integrity. Each dialogue has its own discourse subject but its topical determinateness is sometimes violated by "digressions" from the standpoint of a monologic development of the subject. They are necessary in a dialogue and are something fundamentally different from the topical determinateness of a pedagogical-didactic dialogue whether oral (a lecture, a report and so on) or in a written or printed text (such as a treatise or a textbook). An important role is also played here by the communicants' specific exchange procedures and in particular by the relation of partnership expressed in a so-called Socratic form. The teacher does not allow any hierarchy between the pupil and himself, proclaiming right away: "He knows that he knows nothing". He assumes that the use of the maieutic method will help to bring forth the pupil's latent knowledge. Equally essential are both the teacher's and the pupil's modalities of utterances (inducement, interrogation, narration) in which the process of communication takes place, utterance systems are accumulated and new, already intersubjective, experience is acquired. An important role is also played by specific norms and operations, governing and structuring the synthesis and analysis of utterances by each participant.

When Plato's life reached a turning point, important changes occurred in the teaching system as well as in the methods of transmitting knowledge and in its structuring. There began a process that may be described as a social institutionalization of education: an academy was founded which involved from the outset the integration of a single speech community by recording

---

*A study of oral speech has long been a marginal concern of philology, if not completely neglected. This was due to several reasons, including the subordination of linguistics to philology whose primary goal was a study of written language texts. Now linguistics is beginning to turn its attention to a study of diverse forms of text structure, including the text of oral discourse. Thus Rozhdestvensky singles out three classes of oral speech according to the type of its generation: (1) oral speech with a written prototype-source; (2) oral speech definitely without any such prototype; and (3) oral speech which may (or may not) have a written prototype [20, p. 225].

("reproducing") messages. Plato still had a dual attitude towards written speech. At first, he preferred the dialectics of a natural discourse to written texts. But gradually both the style and structure of his dialogues changed. He was increasingly attracted to other forms of knowledge structuring, to a monologic treatise rather than a dialogue. The very dialogic structure of the knowledge being transmitted receded into the background. It was late in his life after serious changes in the system and ideals of education that he produced dialogues-treatises where the dialogue form was just an appearance, an envelope for the important theoretical content to be transmitted to pupils, supporters and posterity. These dialogues include *Sophist, Parmenides* and *Phyleb*. Some dialogues were first written in narrative form and later rewritten as dialogues. The style of dialogues and Plato's writings, in general, changed. Sometimes they were written in narrative form right away, as treatises. Finally it was Plato's school that produced what could be described as the first textbooks.

The recent intensive studies of Plato's esoteric teaching [30] shows the artificiality of the opposition of his literary dialogues to his oral pedagogical lectures. This schematic division fails to take into account that, firstly, both his dialogic writings and oral instruction had the same source - a Socratic discourse; secondly, that Plato used both forms in philosophical and educational activity alike and, thirdly, that in his seventh letter he developed his theory of a supreme unverbalized intuitive-mystical experience (synopsis), not expressible in either form. Such an opposition blurs the uniqueness not only of Plato's philosophy but of many other schools too (for example, the sophists) as well as of the entire culture of the antiquity that gave priority to a lively dialogue over a literary written text. Plato's propensity for an academic Socratic discourse is manifested, for example, in *Phaedros* [5].

From that time on the narrative form of treatises became predominant although some reflections of oral speech norms could be found both in the writings of Aristotle and in the works of later philosophers. But still the dialogue and verbal communication gradually receded into the background. The status of knowledge structuring and transmission forms changed accordingly. A monologic treatise with its unique composition rules and special semantic characteristics and a textbook where a certain corpus of knowledge was produced by an axiomatic method - such are the ways of structuring knowledge that date back to Plato and became a paradigm for the expression of the knowledge achieved, valid to this day.

Aristotle was precisely the type of scholar who showed preference for the new stylistics - that of written speech, monologic expression of the knowledge achieved and a systematic presentation of one's position. It was this type of structuring that gave knowledge a status of authenticity and validity. But Aristotle would not be a philosopher of the antiquity, if he had excluded from examination the logic of a natural discourse, dialogue and verbal communication [7]. He devotes to these problems his *Topica* where knowledge was assigned the status of probable knowledge. In other words, besides studying the logical structure of apodictical knowledge, Aristotle paid considerable attention to the dialectics of verbal communication, to defining structural forms and functional patterns of knowledge in language communication. What is more, it is probably topological logic, dialectics - in Bakhtin's words, "an abstract product of a dialogue" [10, p. 307] that underlies his logical theory.

In Aristotle's time philosophical treatises, written in narrative form and systematically developing the author's position, performed a dual function - scholarly and educational. As time went on, further institutionalization of education, consolidation of hierarchic distinction and more rigourous regulation of the social roles of a teacher and a pupil produced educational literature proper. Its internal structuring was patterned after scholarly treatises. The typically medieval ritualization of verbal communication, the transformation of a dialogue into a dispute, a strict regulation of communication and even its result completed the canonization and codification of cultural-creative activity. This was primarily reflected in the forms of expounding knowledge, i.e. in writing and then in the other hypostasis of scientific creativity - direct communication between like-minded or differently minded scholars. As a result, oral and written speech proved unable to express the spirit of free search, a distinctive feature of scientific and, indeed any other culturally significant creativity.

The Renaissance opposed all previous form of knowledge transmission, criticized the scholastic systematization of data and regulation of natural communication. As they demolished the former canons of disciplining and revolted against them, the scholars of the Renaissance sought to revive the Socratic mode of knowledge transmission in the context of age-old written culture. These attempts might have proved tragic for the history of culture, had it not been for the emergence of printed texts, a new type of speech and language texts. The sequence of forms of knowledge objectivation and its embodiment in language texts may be presented not only as transition from oral to written speech and then to printed language texts but also as a dual cyclic process.

The first cycle begins with free natural verbal communication, governed by certain norms. Then the dialogue slips to the periphery, and different forms of written speech canonization emerge. The written recording of knowledge gradually leads to the formation of its specific structuring rules whereby the interlocutor seems to fade away from the author's view, the intentional semantic orientation of the text towards another is, as it were, cancelled out and then disappears altogether from the author's conscious awareness. This leads to the emergence of a new type of mentality which Bakhtin aptly termed Ptolemaic - to the transformation of written speech into a self-contained and self-sufficient monologue, developing the author's position in a certain closed system and presupposing no one but a passive listener outside its confines. This cycle ends in a revolt against the disciplining of a searching mind by any canons, and priority is restored to a free scholarly search and open natural dialogue, governed by oral speech norms.

The second cycle originates from the cult of a dialogue, typical of the Renaissance. Gradually, however, natural verbal communication recedes into the background in the value system of classical science, and orientation towards a systematic expounding and monologic expression of genuine knowledge acquires increasing importance. Disinterested cognition of the truth and selfless service to it become the principal value. A confrontation of different points of view is regarded as an exercise, unworthy of genuine knowledge. One of the methodological assumptions of classical science and philosophy is the existence of a universal homogeneous subject that alone guarantees the true and objective nature of knowledge. It is only natural that within this universal homogeneous experience the existence of conflicting viewpoints is regarded as a fact incompatible with a scientific spirit.

Disputes and debates are viewed by nearly all classics of new science and philosophy as a relic of the scholastic period in the history of knowledge and an exercise which is essentially futile. For instance, for Hume disputes are an indicator of an imperfect condition of sciences at the present juncture, showing that something is wrong with them [24, pp. 79-80]. According to one of Kant's major propositions, there is no genuine polemics in the realm of pure reason either [8, p. 629]. He viewed disputes as something gloomy and depressing. The dialogue of natural communication is relegated to the informal domain, to the periphery of scientific activity and takes place in the form of private correspondence, the "invisible college" (to use Boyle's term).

Scientific creativity was looked upon as in internal affair of understanding and penetration into material, coming from outside, whereas in fact it is always extrovert and achieved in communication between scientists. In accordance with these scholarly ideals, the teaching system was based on emphasizing the importance of the pupil's internal work in assimilating an alien world of objectively true knowledge.

The dialogic "Galilean mentality" (Bakhtin) which was the starting point of the science of the Renaissance was gradually replaced by other cultural forms where orientation towards a listener ceases to be a constituent factor. In expounding scientific knowledge in accordance with axiomatic canons, it was purged not only of all controversial problems (thus giving the corpus of knowledge clear-cut and closed outlines) but also of the intentional orientation of the text towards a passive communicant - a listener or a reader. Scholarly texts, "freed" from a specific world view, from a personal touch, and from a specific referential-semantic and value orientation, were to express the depersonalized objectivity of truth by their axiomatic structure. A systemic model of the world was presented in a monologic exposition of knowledge. An objectively true knowledge appeared in a text as an autonomous entity, existing outside individual consciousness.

Translation of spontaneous language communication into the tacit register of perception and passive acquisition of meanings, objectivated in oral or written texts, is the dominant trend in the development of culture in modern history. To use Bakhtin's expression, a "monologic model of the world" [10, p. 306] is taking shape alongside a monotheoretical model of knowledge where a theory, evolved by an axiomatic method, claims to be the sole and most adequate expression of true knowledge.

In accordance with the standards of the "monologic model of the world" and a monotheoretical model of knowledge a specific genre of scientific literature came into being - monographic treatises, intended for a referential-logical and systemic exposition of a single position, for the reconstruction of the world's systemic character and for the explication of the meaning of existence as a self-contained and a self-sufficient whole.*

---

*This applied not only to Newton's treatises upon systems of the world as, for instance his *Mathematical Principles of Natural Philosophy* but also Newtonian *Opticks*, i.e. a theory of certain physical phenomena. The latter was preceded by "Optical lectures" (c. 1728) where the organic fibre of a dialogue is preserved and orientation towards a listener is manifested to a far greater extent. It is precisely these lectures and Newton's correspondence that elucidate the atmosphere in which his conception was taking shape and highlight the internal dialogic structure of genuine scientific creativity, artificially removed from the axiomatically organized *Opticks* (c. 1721).

There are certain correlations between the social and cognitive structures of knowledge. The dialogic form of expression is correlated with the social forms of instruction that arose in the antiquity, the forms of a natural exchange between equal partners in a communicative event. The monologic forms of a message, realized in treatises and textbooks, arose out of the routinization of instruction and are correlated with its new structural forms characterized by a hierarchy of the teacher's and the pupil's social roles, special emphasis on the regulating role of education and upbringing and a strict ranking of the content of teaching material and textbooks. The use during the Renaissance of a dialogue as a way of expounding and transmitting knowledge is correlated to new forms of the social structuring of science, academies, natural communication of equal minds.

SYSTEMS AND NORMS OF DISCIPLINARY KNOWLEDGE

The development of European culture involved the differentiation of the social institutions of education and science, the emergence of specialized social structures for research – learned societies and academies, functionally distinct from universities. It would seem that a study of the education system can hardly throw any light upon such a functionally autonomous institution as science with its own goals, structural forms and ways of reproducing the scientific community. Yet science is a multistructural and polysystemic whole where each preceding form is retained as a basis for the emergence of the next one.* As a multidimensional system, science includes such dimensions as natural verbal communication among like-minded scientists, a monologic message, oral transmission of the knowledge obtained, implying a certain distance between the teacher and the pupil and the embodiment of the results obtained in written and printed texts in a variety of forms from a textbook to a journal publication.

Communication between scientists, an exchange of ideas and scientific achievements can only take place on the basis of universally recognized norms and patterns. Neither mutual understanding nor a common dialogue nor joint research can do without them. Therefore, a semiotic approach to science implies the isolation of a certain group of invariants, functionally relevant to the process of communication. Each type of language texts has its own set of invariant structures.

De Saussure's dichotomy between language as an abstract system of universal norms and a multitude of speech events is too schematic and is substantially modified in modern linguistics, which analyses semiotic invariants of each of the various speech types (oral, written and printed). The studies of Permyakov who isolated certain clichés in the system of oral utterances [14], Grinzer's work on epic language formulas and the linguistic etiquette theory, developed by Rozhdestvensky [20], show that there are certain invariant structures not only at the word level but also in more complex locations [12]. Although these oral speech varieties pertain to folklore texts, the linguistic studies mentioned above are of much greater significance for they bring into focus the isolation of certain invariant normative structures that integrate speech into a single whole, set behavioural models for the participants in verbal communication, discipline its structures both at the level of a single utterance and the level of supraphrasal unities.

---

*It is no coincidence that the Russian word *nauka* (science) is etymologically related to *nauchit'* (teach), i.e. the specific mode of the existence of knowledge in the education system.

Natural verbal communication between the teacher and the pupil in the course of instruction does not have the same norms as oral folklore or epics. These two varieties of oral speech differ in the role of individual initiative, the inventiveness of the creator and the degree of his subordination to the censorship of the group. The antiquity's form of dialogic communication is an encounter of equal minds, an overlap of personalities, opposing their subjectivity and freedom to traditional norms of folklore-mythological mentality. Therefore, the individual mentality of each participant in the dialogue is reflective and critical in relation to the established norms: while questioning their universality, it calls for a new mode of their justification - in the individual's experience and self-awareness.

The semiotic approach to a study of communication forms makes it possible to find ways of describing specific verbal communication patterns, basic not only to the initial stages of knowledge transmission but also to the following stages. The notion that verbal communication is spontaneous and is a monologue on the speaker's part is now regarded as unproductive and narrow. Communication, viewed as the mutual activity of both sides is based on certain normative and regulative structures. And in as much as there is a variety of its types, superposed, as it were, upon one another, these invariant structures are stratified. There is a "build-up" of invariant normative characteristics, assigning to each communicant a language of communication, a system of the categorial segmentation of the reality and a system of norms and standards, governing behaviour and attitude to the reality. The disciplinary segmentation of the entire body of knowledge is linked precisely with one of such invariant structural levels, performing an important normative-regulative function both in training a new generation of researchers and in the very process of research. In disciplinary knowledge characteristics, functioning in various scientific communication events, become "already established" norms.

In terms of its disciplinary image, the "life of science" is regarded as an objective existence of stable cognitive structures, regularly applied norms and means of transforming elements of knowledge structure into a functionally determinate and meaningful component, into something used to promote mutual understanding among scientists, the unity of their cognitive attitudes and conceptual segmentations. A disciplinary image of science as a historically determinate form of reflection upon it deals with just one aspect of scientific knowledge and one of the levels of the complex system of science, i.e. the existence in it of some norms, regulatives and decision standards, stereotyped and stable cognitive structures. It is they that form the foundation upon which the onward movement of scientific knowledge rests.

If the importance of this aspect of scientific knowledge is rendered absolute, the knowledge achieved is dogmatized and transformed into onedimensional, formulaic and cliché-ridden language, a far cry from the language of scientific communication and creativity. Given this approach, the rules of scientific communication and the system of disciplinary knowledge are embodied in an abstract system of self-identical norms. The typological way of thinking which determined the development of the logical-philosophical analysis of science until the mid-nineteenth century amounted to attributing to a system of disciplinary norms an independent existence prior to, and outside, specific cognitive acts and communications. It is precisely the hypostatization of norms and regulative cognitive activity and the universalization of some disciplinary language (especially the language of theoretical physics) that constitute an inalienable feature of all conceptions of science, developed in classical bourgeois philosophy. The disciplinary image of science takes shape at a time when there exists a monotheoretical ideal of scholarliness and when methodological reflection emphasizes comprehension of norms of expounding knowledge and structuring it as a systemic whole rather than a study of the norms of natural verbal communication and the logic of a debate.

The intentional dialogic structure of scientific knowledge and language, the polyphony and polychromy of scientific creativity are not embodied in the methodological mentality of science in modern times. The disciplinary image, arising in the conditions of university science, i.e. the functioning of knowledge in the instruction system, presupposes the restriction of the subject-matter and awareness of the pupil's limited abilities. This image of science is adopted as a pattern for structuring scientific knowledge. The monologic model of scientific knowledge and the physicalist model of the world system are adopted as a way of structuring the entire corpus of scientific knowledge. Physics with its internal structural forms and its specific research rules is made the standard. An assessment of the entire body of knowledge in terms of this standard involves the introduction of some hierarchy into that body, the singling out, for instance, of basic and derived disciplines, disciplines, differing in the degree of embodiment of accepted exposition norms, in their validity and truth.

The disciplinary image in classical science becomes a pattern for the assessment of the entire corpus of knowledge and a way of its ontological justification. As a result, the entire body of knowledge is segmented into many disciplines, differing from one another in their subject-matter, methodological procedures, conceptual formalism, explanatory patterns and models as well as the style of setting out the results. The disciplinary image of science is essentialistic and implies the segmentation of certain domains of the reality, "entities", "substances" or their "attributes" as an ontological foundation of various segments of scientific knowledge.

Given this approach, the major effort of methodological thought is directed toward a search for the norms of the formal congruence of knowledge and various types of its systemic structuring. The existence of a textbook becomes a major criterion of a scientific discipline. Thus both in everyday and scientific thinking scientific and pedagogical disciplines are implicitly identified with each other while canons of scientific exposition of knowledge are transformed into canons of pedagogical-didactic exposition. The exposition of knowledge in a textbook is of a specific character: it should set out only what is known to be true while all controversial questions are left out. A textbook cannot and should not outline ways and means of achieving true knowledge or discuss procedures that led or lead to misconceptions or erroneous results. A description of conflicting theories and research programmes, vying with one another in search of truth, would be out of place in it. The scientific content of a textbook must be substantially processed and hierarchically structured by gradual ascent from the simple to the complex.

The disciplinary image of science is based on the assumption that not only scientific and pedagogical disciplines are identical in their content and in their structure but also that a disciplinary unit of scientific knowledge is identical with theory. In spite of the assumption that a single theory may represent a scientific discipline, classical thinking did not rule out the possibility of this theory being applied to new domains of reality and of extending the sphere of its application. Yet the disciplinary structure of a certain segment of scientific knowledge was predetermined by this theory alone. The disciplinary image of science was moulded as a monotheoretical model. Therefore, segmentation on its basis of the totality of scientific knowledge is inherent in the identification of a system with a norm, of potential models with those already realized, of functional potentialities with mandatory realizations. It was because this image of science was built up for the solution of specific problems of education or teaching that the function of disciplinary knowledge was interpreted solely as normative. Disciplinary knowledge included only what had already been accepted by the scientific community as a norm and a pattern. As part of the disciplinary knowledge to be handed down to the next generations, this "normative" layer turned out to be doubly normative: it became a norm not only for the scientific but also for the "teaching" community.

Meanwhile, contemporary linguistics has already proposed concepts making it possible to go beyond the overly schematic *langue-parole* dichotomy and excessively normative treatment of language, emphasizing its systems nature as well as its modelling function. These concepts were elaborated by E. Coseriu. "A system embraces ideal forms of the realization of a certain language, i.e. the technique and standards for specific language activity; a norm includes models, historically realized with the aid of this technique and these patterns. Thus language dynamics, the way it is formed and, consequently, its ability to go beyond what has already been realized are effected through the system; the norm corresponds to the embodiment of language in traditional forms" [11, p. 175]. The norm involves a choice within the possibilities and variations, offered by the system. According to Coseriu, the language system reveals what is functionally possible, predetermines the coordinate system indicating the open and closed routes while the norm is a system of mandatory realizations. Consequently, the invariants of each of these language structures are different: the norm reflects the universally relevant invariants of realization and the system functionally possible invariants and ideal models of realization. A norm realizes or actualizes what exists in a system as a possibility. Any shift in the system is not something inconsistent with it but, as Tynyanov put it, "a displacement of the system", a change in the norm, creativity on the basis of certain norms.

Such differentiation between a norm and a system may be applied to the analysis of disciplinary knowledge.* One may isolate from it, on the one hand, a system, manifesting functional cognitive potentialities, pre-setting models and technique of research, mapping out guidelines, coordinates and a set of options, and, on the other, the normative realization of the disciplinary cognitive system, the "already realized" choice from the set of possibilities, offered by the system, and the normative realizations available. In this case disciplinary knowledge emerges as a multilevel entity. It will be understood not merely as a totality of normative realizations and the process of their change but as a complex cognitive system, possessing functional possibilties and bringing out layers of possible but not yet realized meanings.

The model of a disciplinary structure of knowledge, inherent in classical science and philosophy, is based on the identification of a system with a norm, of the possible with the already realized, of the potential with the normatively actualized. With the emphasis on the invariants of normative realization, the standards and patterns for the solution of problems, the

---

*It should be noted that Levin applied this differentiation to an analysis of the structure of physical theories by isolating a system of theory and a class of theory-sanctioned realizations - norms [12, pp. 237-242].

disciplinary image of science embodies precisely what has been established, has become petrified, universally mandatory and sanctioned by traditional realization. Meanwhile it loses sight of another dimension of knowledge - its open character, the wealth of possibilities, inherent in it, the systemic character of its categorial and methodological apparatus that can in no way be reduced to a single avenue of disciplinary-normative realization.

The "system-norm" dichotomy makes it possible to interpret disciplinary knowledge as a multitude of theories. Each of them may be construed as one of the normative realizations of the cognitive system of a certain scientific discipline. This makes it possible to go beyond not only the monotheoretical orientation of classical philosophical thought but also the relativistic position stressing the pluralism of theories and their incompatibility with one another. A sequence of theories may be regarded as a succession of normative realizations, based on a specific field of possibilities, predetermined by the system of disciplinary knowledge. The differentiation of a system and a norm enables the historical character of normative realizations to be defined and a layer, richer in possibilities realized merely in part, to be seen behind each normative in variant.

In terms of the semiotic approach it is possible to distinguish between innovation and a change in knowledge, inasmuch as disciplinary knowledge is viewed as a social phenomenon, reflected in each cognitive and communicative act. The transformation of an innovation, proposed by a scientist, into a change in disciplinary knowledge and language is effected solely due to its recognition by the scientific community. In other words, affiliation with the scientific community and recognition of an innovation by it prove to be an important constituent characteristic of science. The populationist way of thinking makes it possible to identify the time lag between an innovation and its recognition by the scientific community, to spell out the functional role of disciplinary patterns and norms in this recognition.

Differentiation between innovations and changes in the system of knowledge permits a sociological interpretation of the very mechanism of innovation acceptance. Inasmuch as the transformation of an innovation into a change involves a deliberate choice on the part of the scientific community, this acceptance is a social and sociopsychological process where value orientations, orientation to a reference group, prestige criteria, etc. play an important role.

An innovation, involved in a scientist's individual creativity, a product of his will and mind, is implemented in the context of a given disciplinary language, differentiated in accordance with the specific character of the research programmes, adopted by a given scientific community (a scientific school, research group, etc.). It is the context of the disciplinary language that ensures the transformation of an individual innovation into a scientific change, i.e. into an innovation, recognized by the disciplinary scientific community. An individual innovation is likewise conditioned by the context of natural scientific communication. Therefore, each innovation must be accepted not only in the referential-semantic, but also in the communicative aspect. In other words its relevant features are orientation to another, intentional direction to another participant in communication and research. Disciplinary language may be construed as interference of diverse language of natural communication. It is imbued with semantic intentions and contexts. Since any research as a communicative process is aimed at a search for an answer, research activity is a complex gamut of cognitive acts and includes a query, an anticipation of an answer, assent and objection to other communicants. In other words, the leading edge of science is a field of interaction of many essentially equal minds. It is only a monologic model of the world and a monotheoretical model of knowledge than turn this meeting place of many minds into a onedimensional, homogeneous and universal mind, and the formation of a polyglot differentiated meaning – into a linear ascent to universal structures of knowledge.

In terms of a semiotic approach the structure of comprehension also appears to be quite different. Construed in the previous conceptions of science as awareness achieved by the internal operation of the mind, it is treated in the communicative model of science with the emphasis on the dialectical interconnection and mutual determination of comprehension and answer. Neither of the communicative acts can exist independently of the other.*

It should be noted that a norm differs from its codification. Codification is comprehension, ascertainment and recording of the norm. Such objectivation of a norm in codified texts has historically conditioned forms. A textbook is intended for the codification of the norm and patterns for problem-solving. It "embodies" norms and standards of disciplinary knowledge, accepted by the scientific community. It gives the results of research a mandatory and compulsory force and a canonized and systematized form. An analysis of various textbooks in some scientific discipline over a certain period will reveal not only a time lag between the achievements of the leading edge of science and the textbook content but also a gap between the value-normative characteristics of exploratory search and the disciplinary-normative standards, adopted in the textbooks. Louis de Broglie, for instance, pointed out that teaching and research were brothers-foes: "Research of necessity implies perpetual unrest while teaching as such seeks to establish imperturbable confidence, opposed to unrest" [4, p. 345].

History of natural science methods, an analysis of textbooks and the ways the content of a scientific discipline is presented in them may provide ample data highlighting the change of the ideals of scholarliness, their spread to the near-scientific environment and transformation into the common sense of science. On becoming a subject of instruction, the disciplinary nucleus is petrified in a rigid system of dogmas while the value-normative system of scientific creativity assumes in the textbooks the form of a universally mandatory system of standards and patterns. The norms of science of a certain period are universalized and become the only way of presenting and arranging material in textbooks [16, 19, 25].

Various attempts are now being made to remodel not only the methods of teaching but also the social structure of the education system. A search is under way for new approaches to organizing instructional material and the process of instruction, some of them being in opposition to the disciplinary-normative structure of knowledge and its presentation. They emphasize instruction through free communicative events, not confined to the

---

*History of science may also take into consideration the dialogue between the readers and the author of a certain text, the actualization of scientific texts by the reader who perceives the texts and reflects upon them. This goal of history of science implies also penetration into the "latent parameters" of a scientific text – explicit and implicit polemics, matching the author's text with the concepts of a like-minded circle, etc.

disciplinary structure of knowledge.* Behind this search in educational science is an effort to find new ways of developing the education system in accordance with the real life of contemporary science and the spirit of contemporary scientific exploration. An immense growth of scientific knowledge, the emergence of new borderline disciplines, the growing complexity of the objects of inquiry have resulted not only in the erosion of boundaries between individual sciences and the interpenetration of their methods but also in a search for new forms of specialization, other than disciplinary. "In our time", Vernadsky wrote, "the framework of an individual science, one of those into which scientific knowledge is broken down, cannot accurately define the field of a researcher's scientific thought or describe his research with precision. The problems he is concerned with more and more frequently go beyond the framework of a single well-established science. We specialize in problems rather than in disciplinar sciences" [6, p. 89].

Indeed, our time witnesses the emergence of a new image of science where the entire corpus of disciplinary knowledge appears as a totality of diverse problems and problem situations and scientific creativity - as the formulation and solution of problems. Each of them may be broken down into a subset of goals attained on the basis of the existing theory, accepted as a pattern and a norm. In accordance with this image scientific activity is interpreted as activity governed by certain norms and rules of heuristic search. The problem image of science highlights two crucial lines of research: the need for profound specialization and penetration into the object from different angles, i.e. a multiple approach to the object of study.

The new image of science makes it possible to view science as a unity of cognitive and communicative acts, as an activity implemented in the communication of scientists with one another and with their predecessors. Science, interpreted as a sequence of various acts, is governed by certain norms and patterns. These norms and patterns are embodied in the disciplinary structure of knowledge. The problem view of science is closely associated with its interpretation as a communicative process, materialized in some texts and normative realizations and governed by certain rules and disciplinary standards.

---

*Compare this with the books by Illich [27, 28] criticizing the contemporary education system and proposing a programme of "descholastization" ("un-schooling") of society. According to the author, the school fosters in a pupil a number of myths (institutionalized, standardized and programmed values). It is necessary to deinstitutionalize education and develop it in open forms, not restricting the initiative and enabling the pupils to actively cooperate and communicate with their peers. According to Illich, deinstitutionalized education is formed within "four networks" of knowledge or sources of instruction (tools, knowledge exchange network, communication and information networks). As opposed to the former methods of organizing education, Illich emphasizes autonomous and creative relations among the pupils and individual freedom, realized in communications.

REFERENCES

1. K. Marx and F. Engels, *Collected Works*, 2nd edition, Vol. 20, in Russian.
2. M. M. Bakhtin, *Problems of Literature and Aesthetics*. Moscow, Sov. Pisatel (1975), in Russian.
3. P. G. Bogatyrev, *Problems of Folk Art Theory*. Moscow, Iskusstvo (1971), in Russian.
4. I. de Broglie, *Sur les sentiers de la science*. Paris, Michel (1960).
5. T. V. Vasilyeva, Plato's unwritten philosophy, *Voprosy Filosofii* No. 11 (1977), in Russian.
6. V. P. Vernadsky, *Reflections of a Naturalist*, Book 2. Moscow, Nauka (1977).
7. B. S. Gryaznov, *On a Historical Interpretation of Aristotle's "Analytics"* Moscow, Nauka (1971), in Russian.
8. I. Kant, *Collected Works*, Vol. 3. Moscow, Mysl' (1967), in Russian.
9. R. Carnap, *Philosophical Foundations of Physics, (An Introduction to the Philosophy of Science)*, (M. Gardner ed.). New York, Basic (1966).
10. *Context 1976*. Moscow,Nauka (1977), in Russian.
11. E. Coseriu, Sincronia, diacronia e historia, in: *El problema del cambio linguistico*. Montevideo, Universidad de la Republica (1958).
12. A. E. Levin, A semiotic approach to the analysis of the structure of physical theories, in: *Problems in the History and Methodology of Scientific Knowledge*. Moscow, Nauka (1976), in Russian.
13. A. P. Ogurtsov, From a principle to a paradigm of activity, in: *Ergonomics*, Issue 10. Moscow, VINITI (1976), in Russian.
14. G. L. Permyakov, *From Proverb to Folk-tale*. Moscow, Nauka (1970), in Russian.
15. Yu. A. Petrov, *Methodological Problems in the Analysis of Scientific Knowledge*. Moscow, MGU (1977), in Russian.
16. V. V. Polovtsev, *Fundamentals of General Methodology of Natural Science*. Moscow, Prosveshcheniye (1914), in Russian.
17. G. Polya, *How to Solve It. A New Aspect of Mathematical Method*. Princeton, Princeton University Press (1946).
18. G. Polya, *Mathematical Discovery*. London (1965).
19. B. E. Raikov, *The Ways and Means of Natural-science Education*. Moscow, Prosveshcheniye (1960), in Russian.
20. Yu. V. Rozhdestvensky, Oral speech patterning, in: *Systems Research Yearbook 1976*. Moscow, Nauka (1977), in Russian.
21. N. Rosianu, *Traditional Folk-tale Formulas*. Moscow, Nauka (1975), in Russian.
22. F. de Saussure, *Papers on Linguistics*. Moscow, Progress (1977), in Russian.
23. *Readings in The History of Russian Linguistics*. Moscow, Prosveshcheniye (1977), in Russian.
24. D. Hume, *Treatise of Human Nature*. London (1899).
25. H. Armstrong, *The Teaching of Scientific Method and Other Papers on Education*. New York (1910).
26. T. S. Eliot, *Ausgewählte Essays, 1917-1947*. Berlin, Hennecke (1950).
27. I. Illich, *Une societé sans ecole*. Paris (1971).
28. I. Illich, *Tools for Conviviality*. New York (1972).
29. K. H. Spinner, Theoretische Pluralismus: Prolegomena zu einer kritischen Methodologie und Theorie des Erkentnisfortschritts, in: *Sozialtheorie und soziale Praxis*. Meisenheim, Albert (1971).
30. *Das Problem der ungeschriebenen Lehre Platons. Beiträge zum Verständniss der platonischen Prinzipienphilosophie*, Darmstadt, Wippern (1972).

SCIENTOMETROLOGICAL INDICATORS OF SOVIET LITERATURE ON SYSTEMS RESEARCH

S. I. Doroshenko

One can consider the mutual laws of science such as the exponential growth of science and the distribution of research results according to the Lotka rule [5] to be adequately valid for a large class of objects. One of the goals of the scientometrology is more to determine specific numeric parameters as applied to the analysis of the dynamics of a data set, the distribution of research results, etc., than to verify these laws. Scientological measurements are generally taken for rather well defined and established areas of science (for recent work see [7]). The aim of this study was to identify and analyse quantitative indices pertinent to the process of storing the set of Soviet literature on an interfacial science like "systems research". This will enable certain conclusions to be made about future functioning of this relatively young field of research. The verification of agreement between these "fuzzy" areas and conventional quantitative laws of science may also be of interest for interdisciplinary research, which is growing in scale as a result of the increasing integration and complexity of the system of scientific knowledge [2].

The very hazy boundary of this area — involving a fairly extensive range of studies from applied research in systems analysis to philosophical problems of systems research — makes it difficult to delineate the set of publications to be analysed. Thus, the study was confined largely to methodological papers on systems research, as it is primarily methodological results that essentially define the future development of both the theoretical and applied aspects of any science. Because there is no accepted established bibliography in this rather young area of research, two indices of publications on systems research, one in 1971-1972 and the other in 1973-1974 were used as a basis for this study. These bibliographies were compiled at the Science Library of Odessa I. I. Mechnikov State University by S. M. Kirichenko, Chief Bibliographer, and Ms I. N. Sarayeva, junior researcher, Laboratory of Systemology and Control Systems, Odessa State University. Bibliographies were also taken from: *The Development and the Essence of Systems Approach* by I. V. Blauberg and E. G. Yudin (Moscow, 1973), and *The Foundations of the General Systems Theory: Logic and Methodological Analysis* by V. N. Sadovsky (Moscow, 1974). These bibliographies are of a high professional standard, so that the sample set based on them and analysed in this chapter can be said to provide a fair representation of Soviet (primarily methodological) literature on systems research.

The sample set consists of papers from journals, books (including collections by different authors), author's summaries of theses and book reviews totalling 1074 publications (277 papers and 170 books among them) by 688 authors. The time period covers the years 1957 to 1974. It was decided to choose 1957 as a reference point because there were relatively few publications on systems research previously, so all these are dated in this paper 1957. The statistical processing included a study of the time evolution of the sample set and of the authors involved, as well as a study of the distribution of the research output rate. The following results were obtained according to these three guidelines.

(1) <u>The time evolution of the publication sample set</u>. The sequence y_t, which is plotted in Fig. 1 (curve 1) with the year 1957 as the starting point, represents the total number of publications vs time t. The function is almost linear, so that the conjecture that the set in question grew exponentially appears quite reasonable Therefore the time series of the sample can be written in the form $y_t = y_0 e^{\nu t}$ where y_0 is number of publications at time t=0 (i.e. in 1957), and λ the intensity parameter describing the increase in the number of publications. It is convenient to rewrite the above relationship in a linear form, i.e. as $\ln y_t = \ln y_0 + \lambda t$ which enables a linear fit for the sample sequence, $\ln t_y$, in Fig. 1 (curve 1), provided that λ can be estimated.

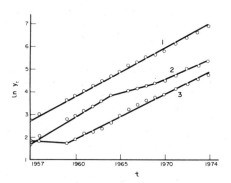

Fig. 1. Semilogarithmic plot of time evolution of the publication set.
1 - general set, 2 - paper set, 3 - book set.

To estimate λ, one uses the sequence of λ_t's, i.e. the sampled values of λ at times t. It can be quite easily shown from the exponential law that this sequence follows the relationship $\lambda_t = \ln (y_{t+1}/y_t)$. λ is taken as the sample set average, i.e.

$$\lambda \equiv \lambda_{av} \equiv \sum_t \lambda_t / n$$

where n is number of samples. The doubling time η, that is the time period over which the set doubles its value, can also be found from the exponential law to be $\eta = \ln 2/\lambda = 0.7/\lambda$.

Statistical processing of the above sample set (see Table 1) with n = 17 resulted in $\lambda = 0.25$ and $\eta = 2.8 \approx 3$ years. This interesting result indicates that in this case the growth rate of the sample set differs greatly from what can be found elsewhere. It takes from 7 to 12 years for a sample set to double in the "mature" natural sciences, while the doubling time is less than 3 years in systems research. This comparison suggests that the area of systems research is "young" and developing rapidly.

Table 1. Time Evolution of Publication Set and Number of Authors

Year	General publication sample	Journal papers	Books	Number of authors with publications this year	New authors	Departing authors	Net number of authors
1957	17	6	7	15	15	0	15
1958	21	10	7	3	3	0	18
1959	31	14	7	10	9	4	23
1960	40	17	9	9	6	3	26
1961	50	23	11	8	3	0	29
1962	64	30	13	12	7	2	34
1963	79	43	14	14	11	3	42
1964	111	53	20	28	22	16	48
1965	142	60	29	37	25	12	61
1966	189	76	40	44	27	20	68
1967	234	81	51	28	18	15	71
1968	280	90	63	44	28	16	83
1969	329	106	73	47	22	19	86
1970	395	127	83	51	21	19	88
1971	551	148	102	192	96	76	108
1972	699	200	123	172	121	130	99
1973	874	252	145	166	118	120	97
1974	1074	277	170	195	136	195	

Typical carriers of scientific information that ensure communications and eventually determine the development of the area are papers in periodicals and books. (These are generally available in libraries only, while a book review cannot be considered to be essentially a piece of research.) The respective subsets were therefore singled out for separate analysis.

The sequence y_t' representing the amount of papers prior to time t (curve 2 in Fig. 1) runs in a less smooth pattern than the time evolution of the entire set, and deviates noticeably from a straight line. As a result, no single exponential fits the y_t' sequence. It is convenient to use a piecewise fit by three straight lines with different slopes. This means that the time evolution of the sample set will be represented by three matched exponential laws with different increments λ. Statistical processing of g_t' (Table 1) produces the following results. Initially (over 7 years, 1957-1964), λ is 0.314 and the doubling time η is 2.55 $\simeq$ 2.5 years. Within the middle stage (the 5 years of the 1964-1969 interval) λ is 0.14 and η is 5 years. At the final stage (the 5 years between 1969 and 1974) λ is 0.2 and η is 3.5 years.

This may be interpreted as follows. It should be kept in mind that papers in journals are the most mobile set in scientific information and the time evolution of this set is the truest representation of the research trends and capacity for publication in a given area. Initially, the area is not delineated, and no theoretical or methodological constraints are imposed. This early stage is generally characterized by the rapid accumulation of the number of papers, with the growth rate of this number (as is the case here) being greater than the general accumulation rate of publications, as it is yet too early to write theses or books, and there is nothing to review. Then, at an interemediate stage the area is stabilized ("narrowed" or rather delineated), the relevant domain of publications is established, and problems appearing at the surface somewhat exhaust themselves. The growth rate of publication naturally declines at this stage, and this is represented by an increase in the doubling period of the paper set from 2.5 to 5 years. (The growth rate of the paper set is now behind that of the general publication set since the build-up of the latter is now the consequence of the increasing amount of books, theses, and reviews.) In the context of this paper, the period beginning in 1969 can be of interest, as the journal paper growth rate went up again (the doubling period fell from 5 to 3.5 years). A possible explanation may be that a special annual publication, *Sistemnye Issledovaniya* (*Systems Research*; in Russian) was begun. This (1) opened a new publication channel, and (2) contributed to better communications in the area which, of course, stimulated new works.

Similarly, a statistical analysis of the book set (Table 1) shows that its time evolution since 1959 can adequately be represented by an exponential law with $\lambda = 0.22$ and $\eta = 3.65 \simeq 4$ years (curve 3 in Fig. 1). As one would expect, this means that the dynamics of the book set is more stable than that for the periodical set. (One interesting point, however, is that the paper set growth rate taken over the three stages is equal to the book set growth rate.) The delay of the book set growth rate compared to that of the papers appears to result from the fact that writing a book often takes years, and requires that a certain basis is initially established. A content analysis of the general publication set indicated that its more rapid growth rate than the book set growth rate can be explained by the upsurge of theses over the last two years (there had been essentially no growth of the thesis set previously.

(2) The dynamics of the author set. The following specifications were determined from the general publication set data: the number of authors with publications in year t; the number of new authors, $\Delta x_t'$, having their first work published in year t; the number of "drop-outs", x_t'', i.e. those having their last work published in year t (see the relevant columns in Table 1). Any co-author is taken to be an independent entity. (This also applies to the counting of the research production rate.) It can be seen that both the total number of authors and the number of new authors rises between 1957 and 1970. There is a four-fold jump in 1971, and then the authorship make-up levels off. A possible explanation can (as with the increase in the paper growth rate beginning in 1969) be attributed to the *Sistemnye Issledovaniya*, whose two issues published as of 1971 appear to have greatly contributed to the increase in the number of authors.

The sample $x(t)$ of the population of authors in the systems research area was calculated for every time t on the basis of the time evolution of the arrival and "drop-out" process using the relationship $x(t+1) = x(t) = \Delta x_{t+1}' - x_{t+1}''$. At t=0 (i.e. in 1957), $x(0) = x_0$ is 15. The results calculated are given in the final column of Table 1 and plotted in Fig. 2. It can be seen from the graph that the author population builds up almost linearly, and then stabilizes at 100 authors by 1970 or 1971.

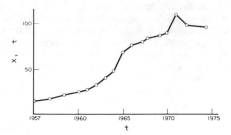

Fig. 2. Time evolution of the size of scientific community.

The evolution of the number of authors was also analysed using the Goffman epidemic model of the development of a research area [6]. Goffman used his model to show that the author "drop-out" rate is approximately constant (possibly zero) whenever the area in question develops steadily, while any abrupt change in that quantity is indicative of a disruption of stability and the onset of an epidemic explosion, an upsurge in the scientific community's interest in this area. The time evolution curve in Fig. 3 shows that the "drop-out" rate is rather constant and approaches zero between 1958 and 1963. Though it becomes nonzero after a transient in 1963-1964, this quantity is essentially constant until 1970, thereby indicating that the area has developed stably prior to that year. The upward jump in the number of departing authors around 1971 and 1972 coincides with the aforementioned upsurge in the net number of authors, which harmonizes with Goffman's theoretical result. It is interesting that Fig. 3 is almost identical to a plot obtained by Goffman from an analysis of a bibliography on mast cell biochemistry research [6].

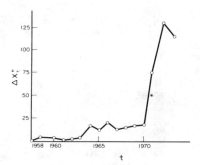

Fig. 3. Time evolution of number of departing authors.

(3) <u>Author production rate distribution</u>, that is, the relationship between the number of authors and the number of their publications, is generally governed by the Lotka law $n_i = Ni$ where n_i is number of authors having i publications, N is number of authors having one piece published, and α is the characteristic factor (usually, $\alpha \simeq 2$). Taking the logarithm of the two sides in the Lotka law, a linear expression is obtained that can more readily be plotted

$$\ln n_i = \ln N - \alpha \ln i.$$

SUBJECT INDEX

Activity
 analysis of 62
 as primordial substance 66
 concept of 62, 63, 64, 66, 67, 69-75, 77, 81, 121, 158
 constructive function 71
 definition of 63
 dialectical materialist concept of 120
 dynamics of 66
 functions of 64
 fundamental role of 71
 Hegelian analysis of 67
 history of study of 76
 Marxist interpretation of 68
 merger with systems approach 63
 nature of 68
 notion of 62
 object-oriented 67, 68, 70, 71, 74
 objective content of 73
 primary object of 122
 principal types of 121
 principle of 69, 72, 73
 psychological analysis of 71, 79
 psychological interpretation of 75
 psychological structure of 70
 psychological theory of 70, 75, 80
 quality of ways and means of 121
 role of principle of 68
 specific projection of 63
 structural presentation of 67, 75
 structure of 74
 systems approach 62-81
 universality of 63, 64
 use of 63
 verbal 76

Activity-action-operation 74
Adequacy of initial assumptions 199-202
Aenads 109
Aggregation level 135, 137, 140
Animal populations 85
Artificial intelligence 36
Artificial systems 39
Artificial (technological) objects of study 39
Automata theory 23, 138-9
Auto-oscillatory phenomena 222
Axiomatic techniques 150

Bacterial cultures 219
Behaviourism 69, 70
Bernoulli's theorem 210
Binary distinguishing procedure 105, 108
Biological objects, state indices of 95-96
Biological systems 23, 213
Biology
 conceptual framework of 91
 systems approach 90-2
 systems research 90-7
 systems theory 82-9
Biomass 219, 220
Black box principle 11
Bottlenecks 75

Capitalism 123
Cell populations 219
Civilization concept 122
Classical thinking 65
Codification 243

Subject Index

Cognition, evolution of 64
Cognitive complexity 38
Cognitive phenomena 207
Cognitive processes 205
Cognitive structures 238, 239
Commitment 161
Communication systems 206
Communications logic 230
Communicative processes 204
Communism 119
Comparability threshold techniques 151
Compensation techniques 150
Competition of hypotheses 222-5
Complex objects
  methodology of studying 92-5
  research 107
  selection of group of elements in 95
  structure and boundaries of 93
Complex problems 148, 149
Complex situation description 46
Complex systems 212, 225, 226
Complexity 37, 38
Components
  distinctive features for 99
  identification 142
  use of term 84
Comprehension, structure of 243
Computer programming languages 142
Computer technology 173
Conceptual components of modelling systems 117
Constitutive distinguishing procedure 104
Context engineering problems 41
Controlled experiment 37
Cost-effectiveness analysis 147, 149-50
Creative and non-creative thinking 58
Cultural norms 163
Cultural phenomena 230
Cybernetics 36, 146

Data acquisition 43
Data aggregation 44
Data banks 44, 192
Data base management techniques 141
Data formalization 43
Data handling 43, 44
Data interpretation 173
Data processing 142
Data storage 43
Data transmission 139
Decision-making 45, 140, 150, 151, 152, 155, 156, 158, 160, 206

Decision options 44
Decision problems 42
Deformalizable components 118
Delay time 140
Descholastization programming 244
Development
  dialectic principle of 11
  principle of 5, 7
Development problems 124
Diagnostic problems 41
Dialectical materialist conception 9
Dialectics 235
Dialogic communication 239
Differential equations 215, 219
Differentiation models 222-5
Diffuse systems 93
Direct description techniques 136
Direct techniques 150
Disciplinary knowledge 229-45
  analysis of 241
  and natural language 231
  content and norms of 233
  semiotic approach to 230-3
  systems and norms of 238-44
Disciplinary language and real-life research and communicative acts 232
Dissipative structures 212
Distinguishing procedures 107
Dyad 104, 109

Ecological models 215-25
Ecological systems 212
Ecosystems 94, 215-17
Education system, social structure of 243
Elements 100, 102, 104, 109, 110
Energy flows 139
Engineering psychology 72, 75
Entities 99
Environment utilization 123
Environmental capacity 220
Epidemic process 215
Epistemological methods 229
Epistemological problems 207
Extensionality concept 101
External factors 202

Feedback 123, 124
Feedforward 123, 124
Files 141, 143
Flow models 217-22
Flow systems 219
Formalization 12, 13
Formalized components 116, 117
Functional structures 11

Subject Index

Functional syndromes 121

Galilean mentality 237
General scientific methodological principles 20
General systems theory 6, 22-5, 52, 91, 146
  as systems metatheory 28
  concept of 25
  methodology of science based on 29
  nature and status of 25
  object 23, 25
  specific features of 7
  theoretical and methodological aspects 27
Generation rhythms 130
Gestalt properties 78
Gestalt psychology 70
Global development, modelling system of 115
Global modelling 115
Global system
  development problems 123
  generalized development alternatives of 124
Gnosceology 9
Goal-attainment mechanisms 159
Goal-orientation 159-65
Goal-oriented programming techniques 45-6
Goal-oriented rational model 157
Goal-seeking behaviour 154-68
  advantages of model 156
  central link in 164
  characterization of 155
  conception of 156
  control limitations of model 156-7
  empirical limitations of model 156
  enlargement of model 159
  existential limitations of model 158
  existing systems model of 155
  methodological limitations of model 157
  theoretical limitations of model 157-8
Goal-seeking-rationality model 156
Goal-seeking systems 11
Goal-setting mechanisms 159-65
Goal-setting systems 11
Gödel's theorem 202
Graphical representation 135, 139
Groups 141, 142

Hierarchic structures 10
Hierarchical simulation systems 45

Hierarchical systems theory 29
History of science 194, 195, 196, 198, 203, 204, 229, 243
Holistic principle in interdisciplinary study of scientific activity 192-210
Human behaviour
  concept of 156
  goal-seeking. *See* Goal-seeking behaviour
  non-optimality of 156
  self-oriented 164
Human objects of study 39

Ideal object 102, 103, 108
Identical self-reproduction concept 199
Incomparability threshold techniques 150
Indirect description techniques 136
Inductive capacities of human intelligence 44
Inductive logic 44
Informal contacts 202
Information documents 193
Information flows 139
Information pattern 138
Information processing 186
Information selection 186
Information service software 138
Information theory 213
Information utilization 43
Instability as prerequisite for development 214
Institutionalization of research 203
Integral hierarchically organized conceptual system 60
Integral object 49, 102-4
Integrative trend 3
Interaction options 42
Interaction structures 11
Interdisciplinary problems 154
Interdisciplinary research 246
Interdisciplinary study of scientific activity 192-210
Interdisciplinary systems approach 8, 12
Internal motives 202
International relations 122
Interpersonal relations 205
Intuition, phenomenon of 57
Invariant structures 229, 230
Inverse problem 10-11
Irrelative definitenesses 100
  conjunction of 100

Journal papers 179

Knowledge
  concept of 65
  disciplinary structure of 244
  model of disciplinary structure of 241
  monotheoretical model of 237
  new 189
  organization of 189
  theory of 229
  transmission of 234
Knowledge generation processes 192, 193, 198-204, 206, 207
Knowledge growth increments 174
Knowledge objectivation forms 233-8
Knowledge representation theory 44
Knowledge transmission 236
Kuhn concept 171

Language concept 65
Language dynamics 241
Least-square fit evaluation 251
Levels of contemporary methodological knowledge 20
Life, systems organization of 91
Life project 160
Limiting factors 224
Linguistics 234
Logic 9
Logical positivism 197, 198
Long-term planning 137
Lotka law 246, 250, 251

Man-machine decision systems 150
Man-machine dialogue facilities 46
Man-machine interactions 150
Man-machine simbiosis 46
Man-machine systems 46, 116
Man-nature interaction 123
Management theory 156
Manuscript rejection 177, 178
Marginality 65, 73
Marxist-Leninist philosophy 9, 120
Marxist theory 5, 22, 49, 67, 68, 107, 115, 118, 119, 120, 121, 124
Material flows 139
Material production modelling 134
Materialism 68
Materialist dialectics 3-15
  and the philosophic systems principle 3-8
  theoretical basis of 7
Mathematical ecology 222
Mathematical modelling 96
Mathematical sociology 214
Mathematical theory 106
Maxwell's equations 203

Measurement theory 43
Memory mechanisms 128, 130
Memory update mechanisms 132
Mentality 70
Metalanguages 231
Metamathematics 9
Metasystems 41
Methodology of science 29
Model development 38, 42
Model sensitivity 45
Modelling systems 13
  approaching choice of basic components 120-2
  choice of basic concept 118-20
  conceptual components of 117
  methodology 44
  nonformalized components of 115-26
  of global development 115
  real systems 145
  set of problems and alternatives of global development 122-4
Monod integral 221
Monographs 182-4, 189
Monologic model of the world 237
Monte Carlo techniques 136
Moral commitment 161
Moral norms 163
Morality, obligatory nature of 161
Morphological analysis 44
Motivation 69, 72
Motivational energy 216
Motive-objective-condition sequence 70
Motive-purpose-condition 74
Myth records 195
Myth study 195
Mythological scale 130

Nature, concept of 66
Negative exponential curve 212
Non-directed activity 165
Non-directed behaviour 165
Nonformalized components 116
  general topology of 116
  in modelling systems 115-26
Non-goal-oriented behaviour 156
Non-linear processes 212
Non-traditional research ventures 35-41
Normal science 199-202, 213, 214, 225
Normative behaviour 163
Norms 163, 164

Object conjunction 101
Object description approaches 134-6

Subject Index

Object generation 101
Object-orientedness 80
Object-transforming procedure 106
Objective structuring 45
Objectivity 10, 12
Obligation concept 161
Observational methodology 197
Open systems
  and development of science 213-15
  evolution of 211-12
  thermodynamic theory of 213, 225
Open systems theory 23
Operations research 36, 147
Opposites 105
Optical lectures 237
Optimization 13
Options comparison techniques 149-51
Oral folklore 239
Oral speech 234, 235, 236, 238
Order concept 86
Organization, definition 87
Organization levels 88
Organizational design problems 43

Pairs of definitenesses 105
Paradigms
  concept of 193-96
  multiplicity and incommensurability of 202
Parmenides dialogues 235
Part and whole, philosophical categories of 49
Performance index 13
Periodic system 78
Philology 234
Philosophical methodology 20
Philosophy 9
Phyleb dialogues 235
Planning 45
Planning model design problems 42
Politico-economic research 193
Power relations 156
Problem area diagnostics 42
Problem area operation 218
Problem situation, external and internal views 9
Problem solving 85, 86, 149, 152
Process description 135
Production processes 134
Programming 45
Psychological systems 23
Psychology
  cultural-historic concept in 71
  cultural-historic school of 72
Ptolemaic mentality 236
Publications. *See* Scientific publications

Purposefulness 37, 40, 47
Puzzle-solving 205

Reality 74, 76
Records 142
Reductionism 70, 192
Reliability assessment 212
Reliability theory 212
Renaissance 236, 237
Reproductive system modelling 127-33
  cultural programme 129
  optimization programme 128
  programme types 128
  strategy selection 128
Research directions 33
Research discipline 36
Research methods and techniques 21
Research ventures 33
Review and collection techniques 44
Review articles 181-2
Reward 157-8, 161

Science
  concepts of 189
  cyclic development of 213
  definitions of 55
  development as open system 211-28
  development of 201, 224, 226
  disciplinary image in 239, 240, 242
  dynamic characteristics of 211
  evolution of 211-15
  history of 229, 243
  life of 239
  limiting factors in 217
  logical-methodological structure of 230
  logical-philosophical analysis of 239
  new image of 244
  normative-value structure of 230
  semiotic approach to 230
  stratum of 230
  study of 72
  trends in studies of 229
Science discipline system 171-91
Science of science 30
Science of science research 172
Science publications 171-91, 207
  echelons of 174-85
  functions of 186-90
  growth rate of 249
  quantitative characteristics of echelons 176
Scientific change concept 196-9
Scientific cognition 60
Scientific communications 229-45

Scientific community 203-6, 213, 215-25, 238
Scientific creativity 236, 237
Scientific discipline 36, 43, 172
  basic processes within 172
  wholeness of 172, 173
Scientific information, typical carriers of 248
Scientific knowledge 60, 68, 230
  dialogic structure of 240
  dynamics of 207
  monologic model of 240
  universals of 229
Scientific literature 173, 237
Scientific orthodoxy 201, 202, 205
Scientific papers 177-81
  time evolution of 248
Scientific problems 73
Scientific progress 33
Scientific revolutions 225
Scientological studies 229
Scientometric studies 17
Scientometrological indicators of systems research 246-52
Self-actualization 165
Self-appraisal 165
Self-assertion 165
Self-control 175
Self-creativity 165
Self-expression 165
Self-knowledge 164
Self-learning system 128
Self-organization 225
Self-regulation 225
Semiotic approach to disciplinary knowledge 230-3
Set theory 51, 55, 98-101, 103
Short-term planning 137
Simplification theory 29
Simulation models 42, 44, 45, 137-9
  as system programming 141-3
  building principles 139-41
  structure modification 143-5
Simulation of structural changes 136-9
Simulation systems 137, 143
  development of 145
  real systems modelling 145
Social behaviour 157
Social control mechanisms 201-2
Social demand 163
Social development 119
Social existence 120
Social groupings 203
Social interaction 157
Social mechanisms 157, 159
Social norms 163
Social reality 67

Social relations 122, 156, 205
Social stereotypes 156
Social structures 132, 238
Social times 132
Socialism 119, 123
Socio-cultural systems 128, 132
Socio-economic factors 39
Socio-economic information 44
Socio-economic modelling 134
Socio-economic problems 117, 123
Socio-economic systems 5, 123, 141-3
Socio-economic theory 67
Sociological concepts 205
Sociological phenomena 76
Sociological research 68, 193
Sociophilosophical analysis 71
Socio-political problems 117, 123
Software systems 40
Sophist dialogues 235
Soviet literature on systems research 246-52
Space-concepts of time 132
Specialized systems 23
Species definition 93
Specific-scientific methodology 20
State indices 95-6
  elaboration of 96
Stimulus-response formula 70
Structural analysis 206, 207
Structural changes, simulation analysis 136-9
Structural multilevel mentality 69
Structure
  and its components 87
  definition of 86
  notion of 10
  of various systems 10
Structure-conscious approach 11
Structure levels 88
Structure role 10
Structurization problem 10
Subject-object relations 9
Subjectivity 10, 12
Submodels 45
Subsystems 42
Supermodel development 42
System
  criterion for identifying 85
  definition of 3, 39, 84
  Mesarovic's formulation of 24
  notion of 10
  Uemov's concept of 24
  use of term 148
Systems analysis 26-8
  I type problems 41
  II type problems 42
  III type problems 42
  IV type problems 42

Subject Index

V type problems 43
  action-oriented 47
  as new direction of scientific research 35-41
  as research programme 33-8
  definitions 6
  development needs 43-5
  development problems 41-7
  developments in 34
  didactic impact of 148
  embryonic state of 33-4
  field covered by 26
  future of 40
  history and concept of 26, 27
  leading role of 41
  limitations of 27
  methodological aspects of practical applications 146-53
  methodological problems of 29
  methodology development 46
  objectives of 9
  present position of 36
  problems of 26
  specific features of 7
  state of the art 34
  theoretical foundation 26
  unity of theory and practice in 13
Systems approach 146-9
  activity 62-81
  benefits of 148
  category structure of 21
  comparative analysis of conceptual structures and fundamental principles 28
  concepts of 21, 26, 146
  creation of science of science 30
  definition of 6
  general schema 147-9
  methodology of science based on 29
  omnipotence of 148
  principal condition of 92
  problems of 154
  research directions 151-2
  specific features of 7
  theoretical framework of 90
  use of term 19-20, 148
  usual meaning of 147
Systems behaviour 40, 50
  at micro- and macro-levels 40
Systems classification theory 28
Systems components 39
  classification of 55
Systems concept 83, 84, 146
  and wholeness relation 59-60
  definitions of 51, 53
  diversified meanings of 50
  extensive use of 59
  intrascientific methodological 16
  ontological status of 54
Systems connection between elements 22
Systems dynamics 11, 39
Systems engineering 147
Systems metatheory 23, 28
Systems methodology 91
Systems norm dichotomy 242
Systems organization
  personal 127, 128
  societal 127, 128
Systems oriented thinking 124
Systems principle 4, 5, 22
  dialectical unity of 7
Systems problems 18, 47
Systems property 21, 22
  lower boundary of 50
  specific 50
  upper boundary of 50
Systems quality 119
Systems research 3-15, 18, 22, 24
  application of methodology ideas and principles 29
  basic concepts 82
  basic object 47
  biology 90-7
  definitions 6, 27, 82
  epistemological features 28
  essential problems 27
  expanded subjective factor 9
  general methodology 27
  main task of 3
  methodological papers on 246
  methodological principles 8-13, 55
  methodological role 59
  methodological studies 16
  methodology of 17, 18, 27
  philosophic substantiation 4
  scientometrological indicators 246-52
  Soviet literature 246-52
  specific methods 28
Systems structure
  internal 50
Systems studies 16, 22, 23, 39
  definition of philosophical foundation of 16
  intrascientific theoretical justification of 16
  objects of 47
  operational rules of 94
  specialized 18
  unresolved questions in development of methodology 27-30
Systems theory 23, 175
  application of principles of 18
  basic concepts 84
  biology 82-9

definitions 6
degree of generality of 24
primary goal of 86

Technical papers 178
Technological systems 40
Temporal organization 131
Temporal parameters 130
Textbooks 184-5, 189
Theoretical thinking 65
Time characteristics 132
Time evolution 89
  of book set 249
  of number of departing authors 250
  of publication set 247, 248
  of research production rate 251
  of size of scientific community 250
Time lags before textbook inclusion 185
Time measures 130
Time organization 133
Time organization units 131
Time parameters 132
Time scales 130, 131
Time structuring 130
Trade-off techniques 151
Tradition mechanisms 233
Transportation network 137
Triviality of research results 177
Truth values 197
Turing machine theory 138
Two cultures concept 129

Uncertainty 45
Uniformity concept 37
Unique model 37
Universal structures 229
Universe of derivative objects 108
Universe of integral objects 108

Valve-oriented behaviour 162
Values formation 162
Variety 101, 103
Volterra equations 217-22

Whole ideal object 102, 103, 108
Whole object 98, 103, 107, 110
Wholeness
  as essential feature of every system 51
  as primary attribute of system 52
  definition 87
  intersections of alternatives and
    relationship alternatives 124
  Marxism-Leninism 119
  of scientific discipline 172, 173
  problem of 154
  subject/object 122, 124
Wholeness concept 49-61, 154
  and system concept 59-60
  as starting point for new coordinate system 56
  definitions of 54
  explication of 58
  fundamental role of 52
  fundamental significance of 50
  generalizing role of 54
  higher types 54
  lower boundary 54
  lower types 54
  meanings of 49
  methodological functions of 55, 56
  philosophical aspects 50
Wholeness criteria 53
Wholeness principle 85
Wholeness theory 98-111, 119
Written speech 235, 236